Manual RS

COMFORT,
AIR QUALITY,
AND
EFFICIENCY BY DESIGN

Air Conditioning Contractors of America

1712 New Hampshire Avenue, NW
Washington, DC 20009
Tel: (202) 483-9370
Fax: (202) 332-5293
E-mail: info@acca.org
Website: http://www.acca.org

Acknowledgments

The development and publication of **Manual RS,** a comprehensive guide to the issues, design concepts, system performance characteristics, and design procedures pertaining to residential comfort conditioning systems, was commissioned by the **Air Conditioning Contractors of America (ACCA) Educational Institute**.

The author, Hank Rutkowski, P.E., ACCA Technical Consultant, gratefully acknowledges the help and guidance provided by his fellow ACCA Manual RS Task Committee members in the preparation of this manual:

ACCA Manual RS Task Committee:

- Dick Pasterkamp, Chairman, ACCA Technical Services Task Team, of Pasterkamp Heating and Air Conditioning Company, Denver, Colorado

- Glenn Friedman, Liaison to the ACCA Executive Committee and 1996 ACCA President, of Engineered Air Systems, Richmond, California

- Al Guzik, Energy Management Specialists, Inc., Cleveland, Ohio

- Jim Herritage, Energy Auditors Inc., Mt. Pleasant, South Carolina

- Dick Shaw, Dean, Ferris State University, Big Rapids, Michigan

In addition, the author extends his appreciation to Marion Wright, who produced the computerized art, tables, and drawings for **Manual RS**, and to Mary Ruth Yao, who proofread **Manual RS**.

Introduction

The performance characteristics of residential heating and cooling systems must be compatible with the architectural features of the home, the occupant's use of the home, and the weather patterns that affect the home. When this interlock is made, comfort and efficiency can be delivered at a price that will be justified by the benefit delivered.

This manual discusses the issues that affect system selection decisions and provides guidance regarding the methods and procedures that must be used to execute the design. This presentation includes information about the parameters that define comfort, the factors that affect indoor air quality, zoning issues, the performance of the thermal envelope, systems and system performance characteristics, equipment performance characteristics, equipment selection procedures, and equipment sizing procedures. This generalized guidance is complimented by other ACCA manuals that pertain to specific design procedures, which include **Manual J** (load calculations), **Manual S** (equipment selection), **Manual D** (duct system design), **Manual T** (air distribution devices), and **Manual H** (heat pumps).

It is important to remember that after a system has been installed, any problems created by an inappropriate design concept or an incorrect design procedure will be difficult and expensive to correct. Therefore, a comprehensive engineering effort (on the original design or a replacement system) always is a good investment.

How to Use This Manual
If the reader wishes to become familiar with all the issues that pertain to residential comfort systems, this manual can be read from cover to cover. This manual also can be used for limited study. For example, information pertaining to water-source heat pump systems can be found in Section 5 and Section 6. The following sections should be studied because they contain information that applies to every design.

    Section 1 — Comfort
    Section 2 — Indoor Air Quality
    Section 3 — Zoning
    Section 4 — Thermal Envelope
    Section 11 — Energy and Operating Cost Calculations

The information contained in the remainder of this manual can be used on an ad-hoc basis. These sections include information pertaining to specific types of systems, primary equipment and supplemental devices, winter humidification, filters, controls, and air-system design procedures. The appendices also provide commentary on testing and balancing, vibration and sound, pipe sizing, wiring, and codes.

Table of Contents

Section 1

Comfort

Section 2

Indoor Air Quality

Section 3

Zoning

Section 3 (Continued)

Zoning

Section 4

Thermal Envelope

Section 5

Residential HVAC Systems

Section 6

Primary Equipment

Section 6 (Continued)

Primary Equipment

Section 7

Winter Humidification

Section 8

Filters

Section 9

Controls

Section 10

Air System Design

Section 11

Energy and Operating Cost Calculations

Appendix 1

Glossary

Appendix 2

Replacement and Retrofit

Appendix 3

Provisions for Testing and Balancing the System

Appendix 4

Vibration and Sound Control

Appendix 5

Pipe Sizing

Appendix 6

Venting Gas- and Oil-Heating Equipment

Appendix 7

Power Wiring

Appendix 8

Codes

Section 1
Comfort

In residential applications, occupant comfort is affected by temperature, humidity, air motion, clothing, the temperature of the interior surfaces of the room, and the quality of the indoor air. The purpose of the HVAC system is to bring these factors in balance with the metabolic activity and the respiratory requirements of the occupants.

1-1 Occupied Zone

The HVAC contractor is responsible for maintaining comfort in a space called the "occupied zone." The boundaries of this space are defined by Figure 1-1. This figure shows that the areas close to the walls and ceiling are not part of the occupied zone. These areas are excluded from the occupied zone because it is neither necessary or practical to control the conditions at these locations.

• These spaces may not be used by the occupants.

• These spaces are used to mix conditioned supply air with the room air.

• During the heating season, the spaces near the exterior walls are subject to convective currents that should be neutralized by the HVAC system.

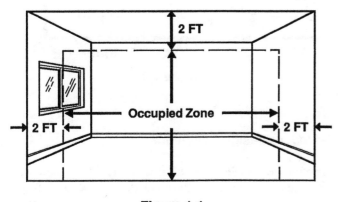

Figure 1-1

1-2 Occupant-Environment Heat Balance

When people lose heat too fast, they feel cold and may shiver. If they lose heat slowly, they get warm and perspire. These heat exchanges occur by convection, evaporation, radiation, and (to a small extent) conduction. In order to insure occupant comfort, these energy transfers must be controlled by the HVAC system and the thermal performance of the building envelope.

Convection
The temperature and velocity of air moving over the skin has a direct effect on the amount of convective heat loss from the skin (or heat gain, if the surrounding air is warmer than the skin). An increase in velocity will always increase the rate of convective heat transfer.

Evaporation
Evaporation creates a cooling effect. The amount of evaporation from the skin depends on the humidity and velocity of the surrounding air. The rate of evaporation decreases as the ambient humidity increases, but an increase in the air motion across the skin will increase the evaporative cooling effect.

Radiation
For radiant heat transfer to occur, the surface of the skin must "see" the radiating surface. If a sight-line is established, exposed skin receives heat energy from any surface that is warmer then the skin and loses heat to any surface that is cooler than the skin. For example, in the winter the inside surface of a large window could be much colder than the skin temperature, causing a noticeable heat loss to the window — even though the room air temperature is normal and body heat loss by convection and evaporation is minimal. (A cold wall could produce the same sensation.)

Structural Considerations
The performance of the thermal envelope obviously has a major effect on comfort. Insulation and double pane windows are important because this type of construction increases the indoor surface temperatures, which in turn reduce radiant heat losses.

For example, on a cold day, the inside surface temperature of a poorly insulated wall might be as low as 61 degrees, compared to the temperature of a fully insulated wall, which might be 71 degrees or warmer. This means that insulation increases the sensation of comfort because exposed skin will radiate less heat to the insulated wall. Since the thermal performance of the building envelope is so important, the HVAC system designer must consider inside surface temperatures — especially the areas associated with glass and poorly insulated walls — when the air distribution system is designed. (This design work should focus on the style, size, and location of the supply air outlets and the effectiveness of the return air path.)

When occupant comfort is evaluated, the temperature of the indoor surfaces are as important as the room air temperature. Tests show that increasing or decreasing these surface temperatures by 1 degree will have the same effect as increasing or decreasing room air temperature by 1 degree.

1-3 Comfort Charts

Figure 1-2 is a reproduction of the comfort chart published by the American Society of Heating, Refrigerating, and Air-Conditioning Engineers (see ASHRAE Standard 55-1981). The shaded areas on this chart summarize the temperature and humidity conditions that are comfortable for at least 80 percent of the occupants during the heating season or cooling season. (Because people dress differently in winter and summer, the indoor conditions that make people comfortable differ slightly between summer and winter. Acclimatization also has something to do with this difference.)

percent RH, depending on structural tightness and the outdoor humidity patterns that characterize the local climate.)

- During the cooling season people have a low tolerance for high humidity. A relative humidity of 60 percent is a suggested upper limit because the percentage of the people that will be comfortable decreases when the indoor humidity rises above this level.

- In cold climates, condensation will occur on the inside surfaces of the window assemblies if the indoor humidity is too high — even if the home is equipped with the most efficient type of windows. To avoid cold weather condensation problems, the indoor humidity should not be allowed to rise above 30 percent RH.

- During the heating season, a change from 20 percent RH to 60 percent RH may go practically unnoticed by most occupants. (As far as comfort is concerned, people are more sensitive to slight changes in air and surface temperatures than they are to changes in humidity.)

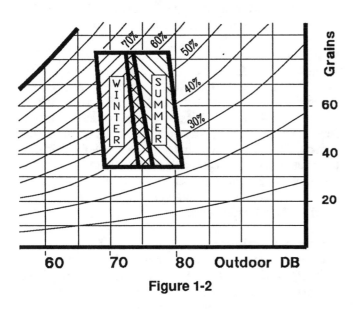

Figure 1-2

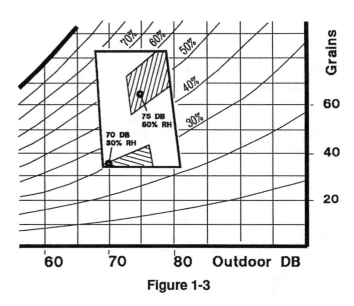

Figure 1-3

However, the size of the summer-winter comfort zones on the ASHRAE comfort chart have some practical limitations. This is demonstrated by Figure 1-3, which provides an adjusted version of the ASHRAE comfort chart. This modified chart suggests comfort zones that are much smaller than the zones on the ASHRAE chart. The differences between the modified chart and the ASHRAE chart are caused by operational limitations associated with the performance of the conditioning equipment and the thermal envelope.

- Except in dry climates, residential cooling equipment is not capable of maintaining a low indoor humidity. (The lower limit could vary from less than 45 percent to more than 50

1-4 Indoor Design Conditions

The modified ASHRAE comfort chart provides a matrix of temperature-humidity combinations that can be used as the indoor design condition. Any of the possible operating points are valid for the comfort of a simple majority, but an increasing number of people will be satisfied as the design condition moves toward the center of the comfort zone. Figure 1-3 also shows the design conditions that have been selected as a basis for the **Manual J** load calculations. These conditions maximize comfort and they provide a reasonable margin of error if the actual operating condition is not coincident with the design point.

1-5 Temperature Swings and Gradients

Residential HVAC systems are subjected to three or four operating loads, depending on whether or not winter humidification is provided. These operating loads correspond to the sensible and latent cooling loads, the heating load, and the winter humidification load. All of these loads can vary from zero to a value that is equal to, or during severe weather greater than, the **Manual J** design load. As these loads increase and decrease, they can cause noticeable fluctuations in the indoor temperature and humidity.

Temperature Near the Thermostat

Residential HVAC systems should control the temperature at the thermostat to within a few degrees of the setpoint during all but the most unusual weather conditions. In this regard, there are three reasons for differences between the temperature near the thermostat and the setpoint.

- During normal operating conditions (loads equal to or less than the **Manual J** design loads), there is always a swing associated with differential between the thermostat make and break temperatures. If an electronic thermostat is used, this swing will be about 2 degrees (-1 to +1), but the swing could be greater (say -2 to +2) if a mechanical device is involved.

- The standard **Manual J** cooling equipment sizing calculations allow for a +3 degree swing during the late afternoon hours of a "design day." (This temperature swing usually does not materialize because the **Manual J** calculations are conservative and because designers tend to "round up" when they estimate loads and size equipment.

- Since the capacity of a properly sized residential HVAC system is not based on the largest possible load, the temperature near the thermostat may drift above (cooling) or below (heating) the normal setpoint during extreme weather conditions. On the average (based on 15 to 20 years of data), this type of weather can occur for about 55 hours during the heating season and 75 hours during the cooling season.

> Residential HVAC systems are not designed to neutralize the largest possible loads because system performance — as related to comfort and operating cost — would be compromised during the thousands of hours associated with intermediate loads. And, when smaller equipment is installed, there are benefits related to installation costs (equipment size and duct size) and utility demand loads (input KW).

Room Temperatures

Industry standards that date back to the 1950s suggest that the temperature in any room should not deviate from the thermo-

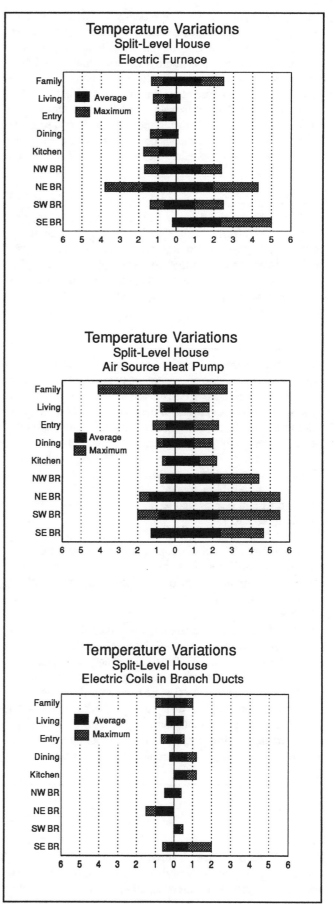

Figure 1-4

stat setpoint by more than 2 or 3 degrees. But, as Figure 1-4 shows, when two types of central single zone systems were tested, the room-to-thermostat temperature differences were as large as 5 degrees. The figure also shows that the room-to-room temperature difference could be as much as 8 degrees. (Even if the temperature at the thermostat is within the 2 to 3 degree requirement, noticeable room-to-room temperature differences can occur. For example, a 4 to 6 degree room-to-room temperature difference is possible if one room is 2 or 3 degrees above the setpoint and the other room is 2 or 3 degrees below the setpoint.)

Figure 1-4 also shows that a multi-zone strategy can significantly reduce the room-to-thermostat and the room-to-room temperature differentials. When this type of system was tested, the room-to-thermostat temperature differences were reduced to the standard differential associated with a standard thermostat, which is about 2 degrees, and the maximum room-to-room temperature differences were held to 3 degrees or less.

Even if the sizing calculations and the installation details are perfect, a single-zone system may not have the ability to maintain close control over the temperature in all the rooms of a home; especially if the home features a complex or multi-level floor plan and/or generous amounts of glazing. However, blower operation on a continuous basis tends to equalize the temperatures throughout the entire house. Unfortunately, two disadvantages are associated with continuous blower operation: one related to the operating cost and one related to humidity control during the cooling season.

• An increase in blower operating hours will translate directly into an increase in the utility bill (about $50 to $150 a year, depending on the size of the blower and the cost of electricity).

• Moisture can be reevaporated into the recirculating air when the compressor is off. This re-humidification effect occurs when the refrigerant coil and the drip pan are wet after the compressor has stopped running. (This is not a problem in a dry climate because the cooling cycle is characterized by a dry-coil operating condition.) Therefore, if marginal humidity control characterizes intermittent blower operation, the indoor humidity may be unacceptably high when the blower operates continuously.

Vertical Temperature Gradients
Floor-to-ceiling temperature differences are normal, but they must not be too large. An acceptable comfort level will be maintained if the temperature difference between the floor

and the ceiling is equal to or less than the differential suggested by Figure 1-5. Note that continuous blower operation (air circulation) may be required to meet this standard because stratification cannot be prevented when the air distribution system is inoperative.

Vertical Temperature Gradients							
Indoor-Outdoor Temperature Difference (°F)							
	10	15	20	30	45	60	75
Heating	0.5	1.0	1.5	2.0	3.0	4.0	5.0
Cooling	3.0	4.5	6.0				

Temperature difference between 4 inches above floor and 84 inches above floor.

Figure 1-5

Level-to-level temperature differences cannot be avoided, but these differences should not be any larger than the room-to-room differences already discussed on the previous page. Therefore, the temperature in the rooms on any level should not deviate from the thermostat setpoint by more than 2 or 3 degrees. (Note that a 4 to 6 degree room-to-room temperature difference is possible if one room is above the setpoint and the other is below the setpoint.)

In general, floor-to-floor temperature control will be noticeably improved by zoning. But, if the home features a single system that serves two or more levels, comfort usually can be improved by operating the blower on a continuous basis.

Even if zoning or continuous fan strategy is used, floor-to-floor comfort may be marginal or even unacceptable if the air distribution system is not properly designed. This means that the air distribution system must feature supply outlets that are carefully located and properly sized, and a low resistance return path for every room that receives supply air.

1-6 Cold Floors

Floor temperatures are important because people do not feel comfortable when their feet and ankles are cold. This particular problem is associated with cold-climate homes that feature slab construction or floors located over unheated spaces. (Floors located over heated basements or heated crawl spaces are not a problem.)

When forced air systems are used to heat homes that have exposed floors, the floor temperature problems can be minimized by installing a perimeter supply air system (outlets located in the floor, blowing up the walls toward the ceiling). Obviously, ceiling outlets and high side-wall outlets are not recommended in these situations because they cannot provide comfort at the floor level. (Overhead air distribution systems are preferred in climates that emphasize comfort during the cooling season.)

Cold floor problems also can be neutralized by baseboard heating fixtures or radiant coils installed below the floor. Usually this approach is found in heating-only homes that are located in cold climates, but this type of equipment can be used to improve the performance of a year-round forced air comfort system. For example, baseboard heat (or below-the-floor coils) could be used to provide comfort in a two-story vestibule area, or baseboard heating fixtures could be used to supplement an overhead supply air system that serves a family room with a slab floor. (A separate control is required for the supplemental heating system and this control should respond to the temperature near the floor.)

1-7 Humidity Control During the Cooling Season

Residential cooling equipment is sized to satisfy a set of design loads that consist of a sensible load and a coincident latent load (see **Manual J**, procedure D). When these loads are imposed on a home, properly sized equipment will operate almost continuously and both of the loads will be completely neutralized. But, this equipment sizing scenario will occur for only a few dozen hours per year. Therefore, for comfort, it is necessary to consider the load-to-capacity relationships associated with the off-peak operating conditions that dominate during the cooling season.

For example, the sensible load will obviously decrease during intermediate weather, but this reduction may not be matched by a proportional reduction in the latent load. This behavior is demonstrated by Figure 1-6, which shows that when the outdoor air is humid, the latent load will remain fairly constant as the sensible load decreases. This means that there will be a corresponding reduction in the sensible heat ratio because the sensible load decreases more than the latent load.

When low-sensible—high-latent operating conditions occur, (which could be for hundreds, or thousands, of hours-per-season), the indoor humidity tends to increase because the cooling equipment does not operate for enough minutes per cycle to neutralize the latent load. This tendency is illustrated by Figure 1-7, which shows that when the compressor run time is abbreviated, the DX coil does not get cold enough to neutralize the latent load. (This figure also shows that the loss of latent capacity is problematic with some types of high efficiency cooling equipment.)

The loss of latent capacity associated with part-load cycling will cause the humidity to drift above the **Manual J** design value, which is acceptable providing that the humidity level stays within the comfort zone. But, if the humidity increase is

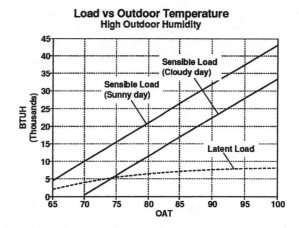

Figure 1-6

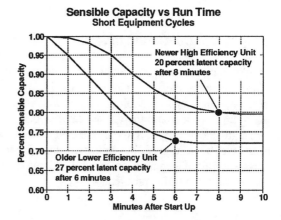

Figure 1-7

too large, the operating point will drift outside of the comfort zone. This loss of control can be minimized by observing the following guidelines.

- If the **Manual J** load calculation is based on a 50 percent RH condition — as opposed to the 55 percent RH condition — the indoor humidity will have more room to drift within the comfort zone.

- The total capacity (sensible plus latent) of the cooling equipment should not exceed the total load (sensible plus latent) by more than 15 percent for cooling-only applications and warm climate heat pump applications, and by more than 25 percent for cold climate heat pump applications.

Comprehensive manufacturer's performance data should be used to size the equipment for the exact set of loads and operating conditions that will be experienced at the site.

The capacity data published in the ARI directory is not adequate because it does not provide information about sensible and latent capacity and because the test-chamber operating temperatures and humidities are not normally the same as the at-the-site operating conditions.

Also note that some climates are too dry to produce a latent load on the DX coil. In this case the indoor humidity depends on the moisture content of the outdoor air, the infiltration rate, and the amount of moisture generated by the occupants. If the outdoor air is dry, these factors will combine to produce an indoor relative humidity that is less than 50 percent, and it could even be lower than 40 percent. But, as long as the relative humidity stays above 30 percent RH, the indoor air condition will be within the ASHRAE comfort zone. (Cooling season humidification is not common, but a humidifier could be installed if the indoor air is considered — for comfort or health reasons — to be too dry.)

1-8 Humidity Control During the Heating Season

During the heating season, very cold weather can produce an indoor humidity condition that is outside of the ASHRAE comfort zone. This dry-air condition causes a sensation of coolness, a desire to increase the thermostat set-point, problems with static electricity, and dry sinuses. Of course, these discomforts can be minimized by adding a humidifier to the heating system. But, if a humidifier is installed, it must not be the source of a condensation problem.

Humidity Required for Comfort
The ASHRAE comfort chart (see Figure 1-2) shows the relationship between comfort, relative humidity, and dry-bulb temperature. This chart shows that acceptable humidity levels can range from less than 30 percent to more than 60 percent, depending on the dry-bulb temperature.

Humidity Required for Health
The ability of the respiratory tract to resist germs and viruses is optimized when the relative humidity is between 30 and 60 percent. Health and comfort problems that are associated with the skin and eyes also can be caused by inadequate humidification.

Humidity Required to Control Static Electricity
A humidity level of 45 percent is required to prevent static buildup on people and on most materials. Somewhat higher levels may be required for wool and some types of synthetic materials.

Upper Limit for Indoor Humidity
The upper limit for the indoor humidity is established by the construction details and the winter design temperatures. Condensation must not be allowed to occur on the inside surfaces of the windows and doors or inside the structural

sandwiches that are associated with the walls, ceilings, and floors. If condensation does occur, the building could be subject to damage from moisture, frost, or ice. Condensation also can cause mold, mildew, and other types of biological growth that can damage the building, produce unpleasant odors, or compromise the health of the occupants.

Surface Condensation
Condensation on the interior surfaces will not occur if the indoor dew point temperature is lower than the temperature of the coldest inside surface. Therefore, the maximum allowable indoor dew point temperature (IDPT) is equal to the temperature of the coldest inside surface (Ts). This temperature depends on the U-value of the structural assembly (usually the window glass), the indoor design temperature (Ti) and the outdoor design temperature (To). The following equation can be used to estimate the maximum allowable indoor dew point temperature. (The psychrometric chart can be used to relate relative humidity, dew point temperature, and dry-bulb temperature.)

$$T_s = T_I - (0.65 \times U \times (T_I - T_o)) = IDPT$$

Concealed Condensation
Concealed condensation can occur whenever the temperature of a surface within a wall, ceiling, or other type of structural component is below the dew point temperature of the air that migrates to that surface. Therefore, if a vapor barrier is not used, or if it is improperly installed, moisture will migrate through the structural assembly; and concealed condensation could occur, even when window condensation is not a problem.

In this regard, the temperature at a concealed surface (Tc) depends on the R-value that is associated with the material between the surface and the outdoors (Rc), the total R-value across the structural component (Rt), the outdoor design temperature (To), and the indoor design temperature (Ti). The following equation can be used to estimate the temperature at a concealed surface.

$$T_c = \frac{T_o + ((T_I - T_o) \times R_c)}{R_t}$$

Vapor Migration
Water vapor rapidly migrates throughout an enclosed space, finding its way into every gap, crack, and cavity that has an air path connection to the space. This means that even though humidification may be done locally, the entire space — as defined by the exterior walls and/or impermeable interior walls — will be humidified. In addition, the HVAC air distribution system can mechanically disperse unwanted moisture throughout a home. (Obviously, impermeable materials, vapor barriers, and an independent HVAC system will be required to control migration from a room that contains a hot tub or a swimming pool.)

Humidification Load

Infiltration and mechanical ventilation cause a humidification load when dry outdoor air displaces an equal amount of humidified indoor air. Another type of indoor-outdoor humidification load is produced when no vapor barrier is installed on one side of permeable building materials. In this case, the vapor pressure difference across the structural components causes moisture to migrate through the porous materials. Internal moisture gains also must be considered. In this regard moisture is added to the conditioned space by the occupants, washing and cleaning activities, cooking, unventilated combustion equipment, plants, fish tanks, and so forth. Therefore the design value for the humidification load is equal to the sum of the various types of humidification loads. Of course, an internal gain is a negative load, and the ventilation load will be equal to zero if there is no provision for mechanical ventilation, and the vapor transmission load can be controlled by installing an adequate vapor retarder.

1-9 Air Motion Within the Occupied Zone

People are very sensitive to air motion, especially when they are seated at rest. Some air motion is good because stagnate air is unpleasant, but too much air motion will be correctly identified as a draft.

Drafts

A draft is created when the ambient air has a combination of velocity and temperature that causes a sensation of coolness and discomfort. The velocity associated with a draft depends on which part of the body is exposed to the flow of air. For example, Figure 1-8 shows the comfort zones that are associated with the neck and ankles. (Note that the neck is more sensitive to air motion than the ankles.)

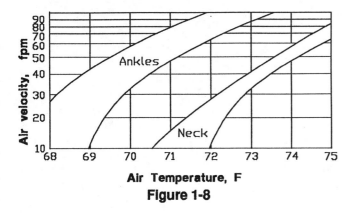

Air Temperature, F
Figure 1-8

The air motion in the occupied zone depends on the natural convection associated with windows and exterior walls and the forced circulation created by the supply air outlets or baseboard heating elements. Complaints about drafts can be expected if these driving forces are not carefully controlled.

- Ample insulation in the walls and thermally efficient windows minimize drafts caused by natural convection. (This is especially important in cold climates.)

- Supply outlets located in the floor, blowing air up the wall, will neutralize the downward flows that are associated with cold glass and cold exterior walls. (Below the window locations are preferred.)

- The jet of air that is discharged from a supply air outlet must never be projected into the occupied zone. The flow from the supply outlet air must be mixed with the room air in the areas near the walls and ceiling (which are outside of the occupied zone).

Stagnate Air

Complaints about stagnate air are caused by insufficient air motion. As it pertains to human perception, stagnate air is associated with velocities that are less than 15 feet per minute. If the supply air is properly and efficiently mixed with the room air (outside of the occupied zone), the air circulation within any part of the occupied zone will be equal to or greater than 15 feet per minute.

Ideal Amount of Air Motion

For occupant comfort, a continuous but imperceptible circulation of air within the occupied zone is desirable. Studies indicate that most people are satisfied when the air circulates through the occupied zone at a velocity of about 25 feet per minute. (Supply outlets that are thoughtfully positioned and carefully sized will produce the desired air motion; provided that a low-resistance return path is associated with every supply fixture.)

> As a point of reference; cigarette smoke hangs in the air when the velocity is less than 15 feet per minute, a sheet of paper might blow off a desk when the velocity exceeds 65 feet per minute, and air should be discharged from a supply outlet at 600 to 700 feet per minute.

1-10 Air Quality

The quality of the indoor air is characterized by its temperature and humidity, the air motion within the occupied zone, and amount of outdoor air that is mixed with it (assuming that the outdoor air is of good quality). In addition, it must be free of fibers, particles, odors, smoke, pollen, spores, biological contaminates, combustion gases, and radon gas. Obviously, it is possible to encounter air quality problems that have nothing to do with the HVAC system. But, if there are no unmanageable sources of contamination (indoor or outdoor), the quality of the indoor air can (and should) be completely controlled by the comfort conditioning system. (Refer to Section 2 for more information about the relationships between the HVAC system performance and the indoor air quality.)

1-11 Noise

Noise is a comfort issue because it affects the occupants sensibilities; therefore, the HVAC system must operate quietly. Noise problems can be avoided by thoughtfully placing equipment, observing duct sizing and installation procedures, properly sizing the supply air fixtures, locating balancing dampers at positions that are remote from the supply outlet, and using duct materials that absorb or attenuate transmitted or generated noise (see Appendix 4).

Residential Single-Zone Systems		
Minimum Performance Standards for Comfort and Safety		
Comfort Item	**Heating Season**	**Cooling Season**
Thermostat setpoint	70 °F	75 °F
Maximum relative humidity	Humidification is optional ... do not exceed 30 percent RH	Maximum of 55 percent RH at **Manual J** design conditions
Minimum relative humidity	Humidification is optional ... 25 to 30 percent RH is desirable	25 to 55 percent RH at **Manual J** design conditions (humidification optional in very dry climates)
Dry-bulb temperature at the thermostat	Thermostat setpoint temperature plus or minus 2 °F	Thermostat setpoint temperature plus or minus 3 °F
Dry-bulb temperature in any conditioned room	Setpoint temperature at thermostat plus or minus 2 °F	Setpoint temperature at thermostat plus or minus 3 °F
Room-to-room temperature differences	Maximum — 4 °F Average — 2 °F	Maximum — 6 °F Average — 3 °F
Floor-to-floor temperature differences	Maximum — 4 °F Average — 2 °F	Maximum — 6 °F Average — 3 °F
Temperature variation from 4 inches above the floor to 72 inches above the floor	1 °F for each 15 degrees of indoor-outdoor temperature difference	3 °F for each 10 degrees of outdoor-indoor temperature difference
Floor temperature (slab floors or floor over cold space)	With thermostat set at 70 °F, the temperature at 4 inches above the floor surface should not be less than 65 °F (except near the outside walls)	
Air Filtration - Minimum	Standard disposable media filter ... ASHRAE dust spot efficiency 10 to 30 percent	Standard disposable media filter ... ASHRAE dust spot efficiency 10 to 30 percent
Air Filtration - Optional	Electronic or high efficiency filter ... ASHRAE dust spot efficiency 50 to 70 percent	Electronic or high efficiency filter ... ASHRAE dust spot efficiency 50 to 70 percent
Air Filtration - Maximum	Pressure drop across filter must be compatible with blower performance	Pressure drop across filter must be compatible with blower performance
Ventilation (outdoor air introduced into the occupied space)	ASHRAE recommends 0.35 air changes per hour (any combination of infiltration and ventilation)	ASHRAE recommends 0.35 air changes per hour (any combination of infiltration and ventilation)
Air circulation within room	Size and location of supply outlets selected for optimum heating performance ... low resistance return path required for every room	Size and location of supply outlets selected for optimum cooling performance ... low resistance return path required for every room
Duct leakage	All ducts and equipment cabinets located in an unconditioned space must be tightly sealed	All ducts and equipment cabinets located in an unconditioned space must be tightly sealed
Exhaust system CFM	50 CFM for baths, 100 CFM for kitchen, all systems dampered	50 CFM for baths, 100 CFM for kitchen, all systems dampered
Flues, vents, and chimneys	See NFPA, GAMA, AGA codes and manufacturer's installation instructions	
Combustion air	See NFPA, GAMA, AGA codes and manufacturer's installation instructions	

Residential Multi-Zone Systems		
Minimum Performance Standards for Comfort and Safety		
Comfort Item	**Heating Season**	**Cooling Season**
Thermostat setpoint	70 °F	75 °F
Maximum relative humidity	Humidification is optional ... do not exceed 30 percent RH	Maximum of 55 percent RH at **Manual J** design conditions
Minimum relative humidity	Humidification is optional ... 25 to 30 percent RH is desirable	25 to 55 percent RH at **Manual J** design conditions (humidification optional in very dry climates)
Dry-bulb temperature at the thermostat	Thermostat setpoint temperature plus or minus 2 °F	Thermostat setpoint temperature plus or minus 2 °F
Dry-bulb temperature in any conditioned room	Setpoint temperature at thermostat plus or minus 2 °F	Setpoint temperature at thermostat plus or minus 2 °F
Room-to-room temperature differences (same setpoints)	Maximum — 4 °F Average — 2 °F	Maximum — 4 °F Average — 2 °F
Floor-to-floor temperature differences	Maximum — 4 °F Average — 2 °F	Maximum — 4 °F Average — 2 °F
Temperature variation from 4 inches above the floor to 72 inches above the floor	1 F for each 15 degrees of indoor-outdoor temperature difference	3 F for each 10 degrees of outdoor-indoor temperature difference
Floor temperature (slab floors or floor over cold space)	With thermostat set at 70 °F, the temperature at 4 inches above the floor surface should not be less than 65 °F (except near the outside walls)	
Air Filtration - Minimum	Standard disposable media filter ... ASHRAE dust spot efficiency 10 to 30 percent	Standard disposable media filter ... ASHRAE dust spot efficiency 10 to 30 percent
Air Filtration - Optional	Electronic or high efficiency filter ... ASHRAE dust spot efficiency 50 to 70 percent	Electronic or high efficiency filter ... ASHRAE dust spot efficiency 50 to 70 percent
Air Filtration - Maximum	Pressure drop across filter must be compatible with blower performance	Pressure drop across filter must be compatible with blower performance
Ventilation (outdoor air introduced into the occupied space)	ASHRAE recommends 0.35 air changes per hour (any combination of infiltration and ventilation)	ASHRAE recommends 0.35 air changes per hour (any combination of infiltration and ventilation)
Air circulation within room	Size and location of supply outlets selected for optimum heating performance ... low resistance return path required for every room	Size and location of supply outlets selected for optimum cooling performance ... low resistance return path required for every room
Duct leakage	All ducts and equipment cabinets located in an unconditioned space must be tightly sealed	All ducts and equipment cabinets located in an unconditioned space must be tightly sealed
Exhaust system CFM	50 CFM for baths, 100 CFM for kitchen, all systems dampered	50 CFM for baths, 100 CFM for kitchen, all systems dampered
Flues, vents, and chimneys	See NFPA, GAMA, AGA codes and manufacturer's installation instructions	
Combustion air	See NFPA, GAMA, AGA codes and manufacturer's installation instructions	

3 1833 04595 1446

Section 2
Indoor Air Quality

Indoor air quality is a complex subject because it depends on the outdoor environment, contaminant-generating capability of the structural materials and furnishings, habits and activities of the occupants, size and location of the structural leakage areas, vents and chimneys, exhaust systems, appliances, performance of the HVAC system, and maintenance practices associated with the structural and mechanical systems. But, as far as the HVAC system is concerned, the indoor air quality only depends on the design concept, the sizing calculations, the installation standards, and the maintenance practices.

2-1 Classification of Air Quality Problems

Figure 2-1 (extracted from the first annual ASHRAE IAQ proceedings, 1992) shows that the majority of the indoor air quality (IAQ) complaints have been classified as ventilation problems. More specifically, these complaints were associated with a shortage of outdoor air, drafts, stagnate indoor air, poor temperature control, and inadequate humidity control.

Also note that a smaller percentage of complaints were traced to biological growth on filters, coils, drip pans, and duct walls, and to condensation within structural components of the home. In other words, more than half of the IAQ problems were caused by an HVAC system that did not satisfy the basic requirements for comfort conditioning.

Figure 2-1 also shows that a significant percentage of IAQ problems are created by sources within the home (smoking, cooking, pets, construction material "off-gasing", carpet "off-gasing", fresh paint, cleaning agents, solvents, and so forth); by pollutants associated with the outdoor air (spores, pollen, dust, and fumes); by sources contained in the soil and ground water (fertilizer, chemicals, radon, and sewer gas); and by biological growth associated with water damage (foundation seepage, plumbing leaks, structural leaks, accidental spills, and natural catastrophes). Obviously, these types of problems are not caused by the comfort conditioning system, but in some cases air-side hardware (exhaust equipment and filtration devices) and/or mechanical ventilation can be used to control the problem.

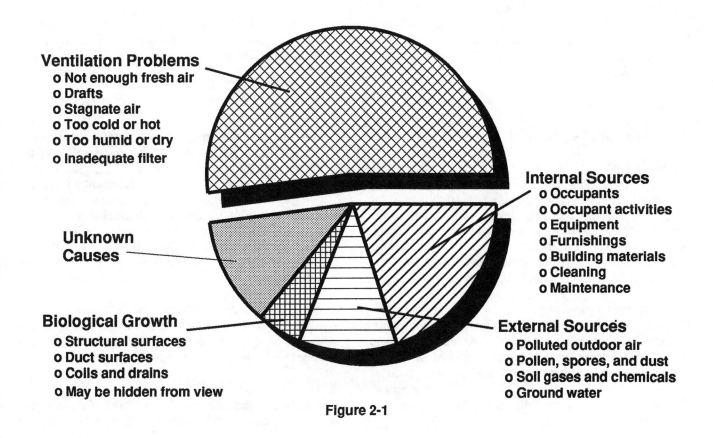

Ventilation Problems
- o Not enough fresh air
- o Drafts
- o Stagnate air
- o Too cold or hot
- o Too humid or dry
- o Inadequate filter

Internal Sources
- o Occupants
- o Occupant activities
- o Equipment
- o Furnishings
- o Building materials
- o Cleaning
- o Maintenance

Unknown Causes

Biological Growth
- o Structural surfaces
- o Duct surfaces
- o Coils and drains
- o May be hidden from view

External Sources
- o Polluted outdoor air
- o Pollen, spores, and dust
- o Soil gases and chemicals
- o Ground water

Figure 2-1

Sources of Contamination

Many different types of volatile organic compounds (VOCs) are used to manufacture laminated and composite building materials (particle board, for example), carpets, furniture, paint, caulking materials, adhesives, and interior finishes. Since these compounds are not chemically inert, they are a source of gaseous contamination, especially when the product is new. (Formaldehyde is an example of a gas that has caused indoor air quality problems.)

- Some information is available about the toxicity and the emission rates associated with VOCs, but this information usually pertains to industrial health hazard applications. Much less is known about the emission rates and concentrations that cause comfort conditioning problems.

- There is not much information about the initial emission rates associated with some common household products and there is a lack of information about how these emission rates decrease over time.

- Most of the toxicity information pertains to a single gaseous compound; little is known about the synergistic effect that is created when two or more compounds act simultaneously.

Housekeeping and maintenance activities can be a source of indoor air quality problems. Projects that involve the use and storage of cleaning agents and maintenance products, (pesticides, cleaning chemicals, paint, sealants, adhesives, and so forth) can generate gaseous pollutants and toxic vapors. Mechanical cleaning (wiping, sweeping, and vacuuming) also can cause air quality problems if these activities dislodge dust, dirt, and biological material that can remain in the air.

Biological growth anywhere within the building envelope can cause air quality and health problems. This type of organic contamination forms on surfaces (visible or concealed) that have been exposed to condensation, leaks, over-spray, and floods.

Air quality problems also can be generated and dispersed by the sanitary systems. For example, complaints are commonly associated with clogged drains, clogged vent stacks, and dry P-traps.

Pesticides, fertilizers, and lawn care products cause air quality problems when the odors, fumes, and chemicals that are associated with these products enter the home with the ventilation air or the infiltration air. Some of these products are very toxic and some may be classified as dangerous or carcinogenic. Even if a product is considered safe, it can cause a serious problem if it is not used in accordance with the manufacturer's instructions. (Most of the information about the toxicity of these products pertain to the threshold limit value that would create an industrial health hazard. Much less is known about the concentrations that are associated with comfort conditioning applications.)

Even mother nature can cause air quality problems. Plants and biological growth produce bioaerosols (dust, fungi, pollen, and spores) that can cause allergic reactions if they enter the home with the ventilation air or the infiltration air.

2-2 Outdoor Air Requirements

Assuming that the outdoor air is of suitable quality, a continuous supply of fresh air should be flowing through the home. This flow is required to dilute the moisture and pollutants that are generated by the occupants, pets, cooking, maintenance projects, and normal housekeeping activities. Outdoor air also is required for fuel-burning equipment that draws combustion air from the indoor space: for example, atmospheric furnaces, gas appliances, fire places and wood stoves.

Fresh Air for Occupants
Outdoor air is required for the comfort and health of the occupants. According to Standard 62-1989 (published by the American Society of Heating, Refrigeration and Air Conditioning Engineers), the fresh air exchange rate should be equal to 0.35 air changes per hour (ACH) or 15 CFM per person, depending on which guideline produces the largest air exchange rate.

- The 0.35 ACH value should be based on the volume of all the rooms associated with the living space, including the kitchen and the bathrooms; but the ASHRAE 62-1989 standard does not specify whether a basement is part of the living space. (If the basement is not included in the living space, the 0.35 ACH value will still translate into a generous amount of outdoor air. For example, based on a three bedroom home that has 2,000 square feet of slab-floor living space, a 0.35 ACH value is equivalent to 93 CFM of outdoor air.)

- According to Standard 62-1989, the number of occupants is equal to the number of bedrooms plus one. (For example, there are four occupants and 60 CFM of outdoor air associated with a three bedroom home.)

The ASHRAE standard furnishes a method for estimating how much outdoor air is required for a home (93 CFM for the

Managing Contaminated Outdoor Air

If the outdoor air is not of suitable quality, the source of the pollution should be identified and removed or controlled. If the source of the pollution is beyond the control of the home owner, the structure could be pressurized by a mechanical ventilation system that features a suitable filter or air cleaning device.

• Infiltration can be reduced, but not necessarily eliminated, when the home is pressured because the rate of infiltration depends on the pressure in the conditioned space, the envelope leakage area, the location of the leakage areas, and the wind velocity.

• The pressurization effect depends on the difference between the mechanical ventilation CFM (outdoor air entering the home through the HVAC equipment) and the exhaust CFM (outdoor air leaving the home through fans and vents).

• The pressure at any specific location within a home is affected by the height of the structure, intermittent exhaust equipment (range hoods, bathroom fans, and clothes dryers), centrally located fuel burning equipment that draws combustion air from within the conditioned space (furnace vents, water heater vents, and fireplace chimneys), and the effectiveness of the return air system (which depends on the number, type, and location of the return grilles and the size of the return ducts).

• The type of filter or air cleaning device will be dictated by the physical characteristics of the pollutant.

• The blower included with a residential air handling package may not have the power that is required to overcome the air-side resistance produced by a high efficiency media filter.

• The ventilation system must be thoughtfully designed, carefully installed, tested, and balanced to achieve the desired performance.

• Some type of air-side heat recovery device might be required to reduce energy consumption and operating costs.

preceding example), but it does not provide explicit guidance regarding the processes and methodologies that can be used to satisfy this requirement. The only instruction on this matter is provided by the following quote:

"The ventilation requirement is normally satisfied by infiltration and natural ventilation."

In other words, the 62-1989 standard stops short of mandating mechanical ventilation when the structure is tightly sealed (infiltration rates below 0.35 ACH). And, it suggests that a combination of infiltration and natural ventilation can be used to satisfy the fresh air requirement. (Natural ventilation includes the movement of outdoor air through open windows and doors, and it includes the crack leakage that is induced by gravity powered vents.) However, there are some problems associated with this suggestion.

• Infiltration and ventilation rates vary depending on the outdoor conditions, the operating mode of the mechanical systems, and the operation of certain types of appliances.

• The flow rates associated with infiltration and natural ventilation cannot be estimated with precision.

• The ventilation air will not be filtered.

• The ventilation air will not be heated.

• The ventilation air will not be cooled or dehumidified.

• Openings and cracks cause drafts (winter and summer).

• Openings and cracks cause cold floors (winter).

• Street noise is transmitted through openings and cracks.

• Open windows and doors reduce privacy and security.

Obviously, natural ventilation is not an acceptable solution to the fresh air problem because it is unpredictable and unreliable, and because it degrades the indoor environment. Therefore, some type of mechanical ventilation system should be considered if the fresh air requirement is not satisfied by infiltration.

Fresh Air for Humidity Control
During cold weather, if a home is too tight, the moisture released inside the home by the occupants, pets, and plants, or generated by occupant activities (cooking, bathing, cleaning, clothes washing, etc.), can cause the dew point temperature of the indoor air to exceed the inside surface temperature of the window assemblies. When this occurs, condensation will form on the glass or frame. Condensation also could form on any poorly insulated structural surface; and if a construction assembly (wall or ceiling) is not protected by a vapor retarder, moisture could accumulate within the structural sandwich. It also is possible for condensation to form inside of a duct run or an equipment cabinet installed in an unconditioned space (an attic for example).

The amount of outdoor air required to avoid cold weather condensation problems varies among homes, depending on the climate, construction details, and the moisture loads produced by the occupants. For a normal amount of occupant-related moisture generation (about 1 pound of water per hour), an infiltration rate of 0.35 ACH will be more than enough to prevent condensation on the inside surfaces of a double-pane window. (The required ACH value could range from 0.15 to more than 0.35, but values in excess of 0.35 would be associated only with an unusually high rate of internally generated moisture.)

Rooms that contain a hot tub or pool must be completely isolated from the rest of the home. This isolation must be structural (vapor barriers and door seals) and mechanical (air leaving a room containing a hot tub or pool should not be returned to the central conditioning equipment).

An occasional episode of window surface condensation is possible if a large amount of moisture is released into the home during a short period of time (by the use of an unvented dishwasher, for example). This probability increases during cold weather when there is no wind-driven infiltration.

The 62-1989 standard calls for a continuous flow of fresh air based on an infiltration rate of 0.35 ACH (or 15 CFM per person), but it does not specify a wind velocity for the infiltration estimate. This lack of guidance is unfortunate because the infiltration calculation could be based on the wind velocity associated with the equipment sizing loads (7-1/2 MPH for cooling or 15 MPH for heating) or some other reference value such as an average seasonal or an average annual velocity.

Until ASHRAE rules on this point, ACCA suggests using the velocity associated with the **Manual J** summer sizing load (7-1/2 MPH) for the infiltration estimate. This means that mechanical ventilation will not be required when the **Manual J** infiltration estimate (cooling) is equal to or greater than 0.35 ACH. This also means that some form of mechanical ventilation should be considered if the **Manual J** infiltration estimate (cooling) is less than 0.35 ACH. (The standard does not insist on mechanical ventilation in these situations, but it would be prudent to provide this feature because infiltration is an undependable source of fresh air.)

Combustion Air and Makeup Air

Atmospheric fuel-burning equipment and appliances must have access to an adequate source of combustion air during any possible operating condition. (Operating conditions vary from moment to moment, depending on the wind; the operation of the HVAC blower; the operation of the bathroom, kitchen, and appliance exhaust systems; the operation of fuel-burning equipment and appliances; the effectiveness of the vents and flues; and, possibly, the position of the interior doors.)

Standard 62-1989

When combustion air or exhaust air is obtained from the conditioned space, the ASHRAE 62-1989 standard (Table 2.3) calls for supplemental ventilation when infiltration cannot satisfy the combustion air or makeup air requirement. The wording as it appears in the standard is provided below.

"Dwellings with tight enclosures may require a source of supplemental ventilation for fuel-burning appliances, including fireplaces and mechanically exhausted appliances."

Unfortunately Standard 62-1989 fails to provide explicit guidance concerning combustion air calculations. However, the standard does contain a footnote that applies to kitchen and bathroom exhaust equipment. This footnote directs the reader to language (contained in paragraph 5.8) that mandates a sufficient supply of combustion air, as indicated below.

"Fuel-burning appliances, including fireplaces located indoors, shall be provided sufficient air for combustion and adequate removal of combustion products."

The standard also demands proof that an adequate supply of combustion air can be provided by infiltration. This text is reproduced below.

"When infiltration supplies all or part of the combustion air, the supply rate of [combustion] air shall be demonstrated."

In Standard 62-1989, paragraph 5.8 also directs the reader to a test procedure (contained in Appendix B) for demonstrating that infiltration will satisfy the combustion air requirement during the most adverse operating conditions (windows and doors closed, no wind, moderate indoor-outdoor temperature difference). This procedure also evaluates the influence of the bathroom and kitchen exhaust fans, the dryer vent, and the fireplace(s).

In other words, some form of ventilation should be considered if there is a possibility that the infiltration rate will be less than the combined combustion-air—exhaust-air requirement. Unfortunately, the standard does not specify how to calculate the combustion-air—exhaust-air requirement, and it does not specify a wind velocity for the infiltration estimate.

National Fuel Gas Code

The National Fuel Gas Code (ANSI Z223.1 and NFPA 54), as published by the American Gas Association (AGA), also addresses the subject of combustion air and ventilation. The following [amended] statements appear in the AGA document.

"When normal infiltration does not provide the necessary [combustion] air, outside air should be introduced [into the space that contains the fuel burning equipment]."

"In addition to the [outside] air needed for combustion, [outside] air shall be supplied for ventilation, including all [outside] air required for comfort."

"Air requirements for the operation of exhaust fans, kitchen ventilation systems, clothes driers, and fireplaces shall be considered in determining the adequacy of an [unvented] space to satisfy the combustion air requirement."

The NFGC does not discuss the ventilation requirements associated with occupant comfort, exhaust systems, household appliances, and fireplaces, but it does provide guidance regarding the outdoor air requirement for primary combustion equipment.

Combustion Air Not Required (NFGC)
According to the NFGC, outdoor air is not required for combustion if the atmospheric fuel-burning equipment is located in an "unconfined space" that does not qualify as "unusually tight construction." The NFGC defines confined spaces, unconfined spaces, and tight construction as follows:

A confined space has less than 50 cubic feet of volume for every 1,000 BTUH of aggregate input heating capacity (atmospheric burners).

Unconfined spaces have more than 50 cubic feet of volume for every 1,000 BTUH of aggregate input heating capacity (atmospheric burners).

• The aggregate input capacity is based on the input ratings of all the fuel-burning equipment and appliances installed in the space. (The heating capacity of direct vent equipment should not be included in this calculation.)

• Rooms that communicate directly with the space that contains the fuel-burning equipment (through openings that are not fitted with doors) can be considered to be part of an unconfined space.

• The volume of rooms that do not communicate directly with a space that contains fuel-burning equipment can be added to the volume of a confined space if two openings (one high and one low) are cut into the wall that separates the confined space and the adjacent space. (Each opening shall have 1 square inch of free area for each 1,000 BTUH of aggregate input heating capacity, but not less than 100 square inches of total area.)

An unusually tight home must have all of the following features. (This type of construction would be equivalent to a

"best" rating as it pertains to the **Manual J** infiltration calculation.)

• Exposed walls and ceilings fitted with a continuous vapor barrier (or an air infiltration barrier).

• Walls and ceilings have gaskets or seals at all of the electrical outlet and lighting fixture openings.

• Tight-fitting windows and doors.

• Sealant applied to the cracks that are associated with the exposed framing (window and door frames, sole plates, headers, etc.).

• Sealant applied to all envelope penetrations (electrical connections, gas piping, refrigerant piping, and so forth).

The NFGC provides a procedure for testing the safety of an existing installation when all the exterior openings (windows and doors) are closed (see Appendix H in the NFGC). This procedure evaluates the influence of the local exhaust systems, household appliance exhausts, interior doors, and the fireplace damper as they affect the pressure in the room that contains the atmospheric combustion equipment.

Combustion Air Required (NFGC)
The NFGC requires a dedicated ventilation system if atmospheric combustion equipment is located in a confined space (which would include any installation associated with "tight construction"). This ventilation system could be driven by natural forces or it could be mechanical.

Natural ventilation
The combustion air requirement can be satisfied by two permanent openings (one high and one low) that freely communicate with a source of outdoor air. (Well-vented attics and crawl spaces qualify as a source of outdoor air.)

• If the openings access the source of outdoor air directly, each opening shall have 1 square inch of free area for each 4,000 BTUH of aggregate input heating capacity.

• If the openings access the source of outdoor air through vertical ducts, each opening shall have 1 square inch of free area for each 4,000 BTUH of aggregate input heating capacity. (The ducts must have the same cross-sectional area as the opening.)

• If the openings access the source of outdoor air through horizontal ducts, each opening shall have 1 square inch of free area for each 2,000 BTUH of aggregate input heating capacity. (The ducts must have the same cross-sectional area as the opening.)

Mechanical Ventilation

The National Fuel Gas Code states that the ventilation requirements can be satisfied by a mechanical system (engineered system), providing that it is approved by an appropriate authority. In this case, the requirements associated with a natural ventilation system (as described above) do not apply.

Contractor Responsibility

No one knows exactly how much outdoor air will be required for a particular home because there are numerous interrelated variables. If the predominant sources of pollution are associated with a few occupants and normal housekeeping activities, the ASHRAE 0.35 ACH could be too generous, especially if combustion air is not required. On the other hand, it is possible to encounter a set of conditions that would not be satisfied by the 0.35 ACH value. Ultimately, the HVAC contractor will have to decide if some form of mechanical ventilation is required (see Section 2-8).

- As a minimum requirement, the structure and the HVAC system must be designed in accordance with local codes and utility regulations. (These documents may refer to ASHRAE Standard 62-1989 or to the National Fuel Gas Code.)

- As a minimum requirement, fuel-burning heating equipment, appliances, and fireplaces must be installed, vented, and adjusted in accordance with the manufacturer's instructions (providing that this direction does not conflict with codes and regulations).

- The NFGC guidance should be followed when this document holds the design to a higher standard than the local codes and regulations.

- A natural or mechanical ventilation system is required for any home that features tight construction (as defined by the NFGC or Table 5, **Manual J** — best) if the home is heated with atmospheric fuel-burning equipment. (Tight construction is indicated by a 50 Pascal blower door test that produces a measured flow equivalent to 5 air changes per hour or less.)

- Combustion air requirements can be eliminated by using electric heating equipment or direct vent furnaces, electric water heaters or direct vent water heaters, and electric appliances.

- Consider mechanical ventilation when the **Manual J** cooling season infiltration estimate produces a value that is less than 0.35 ACH.

- Mechanical ventilation should be considered if a 50 Pascal blower door test produces a measured flow equal to or less than 7 air changes per hour.

- Mechanical ventilation is required when the **Manual J** cooling season infiltration estimate produces a value that does not satisfy the 15 CFM per person outdoor air requirement, as per Standard 62-1989.

- Mechanical ventilation should be installed if a 50 Pascal blower door test produces a measured flow equal to or less than 5 air changes per hour.

- Mechanical ventilation systems must be carefully designed to provide the desired level of safety, comfort, and efficiency.

- The procedures that are found in Appendix H of the NFGC and Appendix B of the 62-1989 Standard should be used to demonstrate that an adequate supply of air is available for atmospheric combustion equipment.

2-3 Exhaust Equipment

ASHRAE Standard 62-1989, Table 2.3 requires ventilation for kitchens and bathrooms. This standard also indicates that this requirement can be satisfied by installing exhaust equipment or by opening windows. Since window ventilation degrades the indoor environment (unfiltered outdoor air, drafts, street noise, and security problems), exhaust systems are preferred. If exhaust systems are installed, they must satisfy the standards performance requirements.

- Kitchens must be equipped with either a 100 CFM intermittent exhaust system or a continuous 25 CFM exhaust system.

- Bathrooms must be equipped with either a 50 CFM intermittent exhaust system or a continuous 20 CFM exhaust system.

- The exhaust air can be obtained from the room containing the exhaust equipment and from adjacent rooms.

- Mechanical ventilation is required if infiltration cannot simultaneously satisfy the exhaust system makeup air requirement and the combustion air requirement.

- Refer to Standard 62-1989, Appendix B, or the NFGC (Appendix H) for procedures that can be used to determine if mechanical ventilation is required.

2-4 Duct Leakage

Duct leakage can affect the indoor air quality if leaks are associated with a duct run located outside of the conditioned space (attic, crawl space, garage, storage, or shop area). Duct leakage can either improve or degrade the quality of the indoor air, depending on the leak's location and the quality of the surrounding air.

Undesirable Effects of Return-Side Leakage

If the leakage is on the return-side of the duct system, the surrounding air will be pulled into the duct system. When this occurs, it can have an undesirable effect on the quality of the indoor air.

• If the captured air is of poor quality (fumes from a garage, storage area, or crawl space, for example) the pollution will be dispersed throughout the home. (Standard filters do not remove odors.)

• If the return air system features filter-grilles and if the captured air is full of dust, pollen, fibers, or spores; particulate material will be dispersed throughout the home (because the leaks are downstream from the filter).

• If the return air system features a central filter, and if the captured air is full of dust, pollen, or spores; particulate material could be dispersed throughout the home if the filter's dust spot efficiency is not compatible with the size of the particles.

• If the return air system features a central filter, and if the captured air is full of dust, pollen, or spores that can be trapped by the filter; the filter loading rate will be increased and more filter changes (or cleanings) will be required.

• If the captured air is relatively hot, the sensible load on the cooling coil will increase and the ability of the cooling equipment to control the indoor temperature could be compromised (depending on the size of the sensible and latent loads and the capacity of the equipment).

• If the captured air is humid, the latent load on the cooling coil will increase and the ability of the cooling equipment to control the indoor temperature and humidity will be compromised (depending on the size of the sensible and latent loads, the capacity of the equipment, and the coil sensible heat ratio).

• If the captured air is relatively cold, the heating load will increase and the ability of the heating equipment to control the indoor temperature could be compromised (depending on the size of the heating load and the capacity of the equipment).

Desirable Effects of Return-Side Leakage

If the leakage is on the return-side of the duct system, the net effect could be desirable if the captured air is directly or indirectly coupled to the outdoors. (This assumes that the captured air is of good quality and that the equipment has adequate cooling and heating capacity.)

• The ventilation rate will be increased and more outdoor air will be available for atmospheric combustion equipment, exhaust equipment, dilution of odors, and heating season humidity control. (*Always check the performance of atmo-spheric combustion equipment — for efficiency and safety — after sealing return-side leaks.*)

• The home will be pressurized and infiltration will be eliminated or reduced. (Infiltration is not filtered and it tends to cause drafts.)

Undesirable Effects of Supply-Side Leakage

If the leakage is on the supply-side of the duct system, conditioned air will be expelled into the unconditioned space. When this occurs, it could have an undesirable effect on the quality of the indoor air.

• The home will be depressurized and infiltration will increase. (Infiltration is not filtered and it can cause drafts.)

• Orders and moisture might be drawn from unconditioned spaces. (An attached garage or crawl space could be at a higher pressure than the conditioned space.)

• Temperature and humidity may drift out of control if the equipment does not have enough capacity to overcome the loss of the conditioned air and the increase in the infiltration loads.

Even if duct leakage improves the quality of the indoor air, it is undesirable because it increases equipment loads, utility demand loads, energy consumption, operating cost, and, possibly, maintenance cost. Any benefit associated with duct leakage can and should be delivered by some type of engineered ventilation system.

2-5 Air Filters

As indicated by Figure 2-1 on the next page, filters rated at 10 percent average dust spot efficiency can capture the larger airborne particles that otherwise would collect on coils and fan blades or cause maintenance and cleaning problems in the conditioned space. However, these filters are not particularly effective at capturing the small airborne particles that cause air quality problems and coat inside surfaces of the air distribution system. (This is the type of filter that is normally supplied with a furnace or air handling cabinet.)

Medium-efficiency and high-efficiency filters can be used to improve air quality, but the amount of improvement will depend on the type and efficiency of the filter. As indicated below, a relatively high dust spot efficiency is required to eliminate the air quality problems associated with very small particles and tobacco smoke.

• Extended surface (pleated) filters that are rated between 25 and 35 percent ASHRAE dust spot efficiency are effective on large biological and mineral particles. These filters can be expected to provide protection only against a buildup of dirt on coils, heat exchangers, and blowers.

- Extended surface filters rated between 35 and 95 percent ASHRAE dust spot efficiency can be used to control a wide range of biological and mineral particles (except viruses and the microscopic smoke and dust particles) with an effectiveness that depends on the efficiency rating. (Small dust and smoke particles cause smudges and stains around supply air outlets.)

- Extended surface filters rated above 95 percent ASHRAE dust spot efficiency are effective on all but the smallest viruses, smoke, and dust particles.

- Electronic filters that have an ASHRAE dust spot efficiency rating that exceeds 60 percent will effectively eliminate particles that cause allergic reactions, smudges, and stains. They will be partially effective at removing biological particles.

- Air quality problems associated with odors and fumes will not be resolved by mechanical filters because they are not effective on gases. If odors are a problem, an air washer or a special filter will be required.

The **dust spot efficiency** rating applies to better quality filters. Do not confuse this with the **dust spot arrestance** rating associated with low-efficiency filters. A high dust spot arrestance rating is not equivalent to a high dust spot efficiency rating.

The efficiency rating of a filter is not the only factor that affects filter selection. When residential air handling equipment is involved, the pressure drop associated with the filter could be a limiting factor if the blower does not have the ability to overcome the resistance that is produced by the filter. In this regard, an electronic filter has an advantage over a high-efficiency media filter because of the relatively small pressure drop that is associated with an electronic filter.

- The pressure drop across a high-efficiency media filter is normally much larger than the pressure drop across a "standard equipment" media filter.

- The pressure drop across an electronic filter is comparable to the pressure drop across a "standard equipment" filter.

The resistance that is added by a filter upgrade must be included in the duct sizing calculations. The **Manual D** Friction Rate Work Sheet will indicate if the blower has enough power to overcome an increase in filter resistance.

Filter Capability			
Filter Design	Upper Dust Spot Efficiency Range	Pressure Drop	Smallest Particles Removed
Standard Viscous Panel	About 10%	Compatible with residential equipment	Hair, pollen spores, soil dust, coal dust
Self-charging Media	About 10%[1]	Compatible with residential equipment	Hair, pollen spores, soil dust, coal dust
Media, Pleated Panel	About 35%	Conditionally compatible with residential equipment	Skin flakes and animal dander; larger grease, smoke, and bacteria particles; larger insecticide and household dust particles
Media, Extended Surface	About 95%	Not normally compatible with residential equipment	Larger viruses; grease and smoke particles; most household dust particles
Electronic Air Cleaner	Above 60%	Compatible with residential equipment	Most bacteria; insecticide dust, largest viruses; larger grease, smoke, and household dust particles
HEPA	Above 95%	Not compatible with residential equipment	Used for clean rooms
1) Efficiency decreases with increase in humidity			

Figure 2-1

2-6 Temperature and Humidity Control

Even though there are no contaminates involved, discomfort that is caused by poor temperature and humidity control is commonly classified as an air quality problem. This type of problem can be avoided if the HVAC system is properly design, installed, controlled, and maintained.

- In some cases, a zoned system may be required to control the temperature in every room.

- If the home features a single-zone system, continuous blower operation may be required to maintain acceptable control over the temperature in every room.

- The central equipment must be precisely sized (using the **Manual J** procedure and comprehensive equipment performance data) to insure adequate temperature and humidity control during any possible operating condition.

- Multi-speed or variable speed equipment improves temperature and humidity control during part-load operation.

- Humidification equipment will improve comfort during very cold weather.

- If the climate is very dry, humidification equipment will improve comfort during the cooling season.

- Humidification equipment must never cause direct wetting or indirect wetting (condensation) of a structural surface (external or hidden), duct wall, or cabinet panel.

2-7 Air Motion in the Occupied Space

Even though no contaminates are involved, the discomfort caused by drafts and stagnate air is frequently categorized as an air quality problem. However, this type of complaint can be avoided if the air distribution hardware (supply air outlets and returns) is suitable for the application. This hardware also must be properly located and correctly sized (refer to **Manual T**).

- Room air motion is controlled by the performance of the supply air outlets.

- A low-resistance return path is required for every room that receives a flow of supply air (except bathrooms, utility rooms, or rooms that contain a hot tub).

- Constant volume systems are less of a problem than variable volume systems because the supply air outlets do not have to accommodate reduced air flow during part-load conditions.

- Vertical discharge through perimeter floor outlets will offset drafts associated with cold vertical surfaces (windows and poorly insulated walls).

- During cold weather, vertical supply air discharge through perimeter floor outlets will provide the most comfort at the floor level when the construction features concrete slab floors or floors over unconditioned space.

- During cold weather, ceiling outlets cannot provide comfort at the floor level when the construction features slab floors or floors over unconditioned space.

2-8 Ventilation Strategies

Mechanical ventilation can be effectively provided by introducing a small amount of outdoor air — normally 50 to 100 CFM, controlled by a hand damper — into the return-side of the central air distribution system. This way, the ventilation air will be fully processed before it enters the conditioned

space, and the home will be marginally pressurized as an equivalent amount of air exits against the resistance of the bathroom and kitchen exhaust systems.

The primary disadvantage associated with this strategy pertains to the loss of ventilating effect when the blower is not operating. Of course, continuous blower operation would solve this problem, but the temperature and humidity of the supply air will not be under precise control when the fuel conversion equipment is inactive. In this regard, the potential for creating a comfort problem depends on the severity of the climate, the percentage of outdoor air, and the amount of excess equipment capacity.

- ASHRAE's suggested ventilation rate (0.35 ACH or 15 CFM per person) translates into 50 to 125 CFM of outdoor air for most homes. (Note that the 0.35 ACH criterion is subject to question as the size of the home increases. For example, a 3,500 square foot home would require about 165 CFM of outdoor air, which would be enough for 11 people. Therefore, unless a code states otherwise, consider using the 15 CFM per person criterion for larger homes.)

- A ventilation rate of 50 to 125 CFM of outdoor air will be about 5 to 10 percent of the blower CFM, depending on the size of the HVAC system. (If a 1,500 CFM system is installed in a 2,500 square foot home, a 125 CFM ventilation rate is equal to 8 percent of the blower CFM.)

- During the heating season, mixing 5 to 10 percent outdoor air with the return air will produce a supply air temperature that ranges between 63 and 70 degrees (10 percent outdoor air) as the outdoor temperature increases from 0 to 70 degrees. This should not cause a problem if the heating plant is properly sized because it will operate almost continuously during the coldest weather.

- During the cooling season, mixing 5 to 10 percent outdoor air with the return air will produce a supply air temperature that ranges between 77 and 75 degrees (10 percent outdoor air) as the outdoor temperature decreases from 95 to 75 degrees. This should not cause a problem if the cooling plant is properly sized because it will operate almost continuously during the hottest weather.

- In a humid climate, mixing 5 to 10 percent outdoor air with return air during the cooling season will add moisture to the indoor air when the compressor is off. This should not cause a problem on hot days, providing that the 50 percent relative humidity option (see Table 1, **Manual J**) is used to calculate the latent load and the equipment operates almost continuously. However, humidity control could a problem during part-load conditions, depending on the moisture content of the outdoor air and the cycling rate of the compressor. (Part-load humidity control in a humid climate is tenuous when a comfort conditioning system has no ventilation capability.)

• If the flow of ventilation air is expelled through a heat recovery device, the operating cost will be reduced. However, during the cooling season, this strategy could acerbate the part-load humidity control problem if the reduction in sensible load is not matched by an equivalent reduction in latent load.

Other ventilating strategies involve using an exhaust-fan system to pull air into the home through envelope openings (natural or engineered) or a supply-fan system to push air out of breaches in the structural envelope (natural or engineered). However, these approaches are not recommended because drafts and comfort problems can be created when unconditioned air is introduced to the occupied space (the occupants' perception would be similar to living in a leaky home). Also note that the exhaust system approach could produce a negative pressure condition that causes back-drafting through vented combustion equipment, and it has the potential to draw contaminated air from ancillary spaces or gases from the soil surrounding a basement wall.

The pressure control problems associated with a supply-fan system or a return-fan system can be avoided by using a push-pull system, which means that two fans (supply and return) will be required. In addition, some type of air distribution system is needed to comfortably mix the ventilation air with the room air. However, this strategy may not be physically or economically compatible with a home equipped with a forced air conditioning system, because two distribution systems must be installed in the same space. (Dedicated ventilation systems are more compatible with hot water heat, electric baseboard heat, or radiant heat.) Also, it is hard to justify this type of system unless a code requires a substantial amount of ventilation air — much more than the 50 to 125 CFM mentioned above. In addition, the benefit and cost associated with installing a heat recovery device should be evaluated.

2-9 Maintenance and Installation

Even if the HVAC system has performance characteristics that are perfectly suited to the home, air quality problems could occur if the equipment is not properly maintained. In some cases, the maintenance procedures themselves could be a source of an air quality problem. And, incorrect installation practices, particularly those associated with humidification equipment, atmospheric combustion equipment, vents, drains, or duct sealing can cause air quality problems.

• The amount of particulate material that is introduced into the conditioned space increases when filter efficiency is degraded by careless installation procedures and negligent maintenance practices.

• Dust-loaded filters reduce system air flow, degrade comfort, and cause the air-side surfaces (duct walls, coils, fans, supply outlets, etc.) of the distribution system to become coated with dirt.

• If return-side duct leakage is associated with an unconditioned space, odors and particulate contamination could be pulled into the duct and distributed to the conditioned space.

• If air-handler panel leakage is associated with an unconditioned space, odors and particulate contamination could be captured and introduced into the conditioned space.

• Dirty coils (refrigerant or electric) can be a source of odors, particularly if biological material is involved.

• Comfort is compromised if the refrigerant system does not have the correct charge of refrigerant.

• Comfort could be affected if the operating controls and the operated devices (flow control dampers, for example) are not properly adjusted.

• Health and safety is at risk if the vents are improperly sized or incorrectly installed.

• Health and safety is at risk if combustion controls and burners are not properly adjusted and maintained.

• Odors and health problems are created when biological growth accumulates on poorly maintained drip pans, heating coils, and refrigerant coils.

• Biological growth also is associated with improperly installed humidifiers, surface condensation (on ducts, pipes, heat exchangers, or equipment cabinets), hidden condensation (inside or under insulation materials), and water-damaged material.

• Air quality problems can be caused by the HVAC system maintenance procedures if the chemicals associated with the maintenance operations produce a lingering source of odors and fumes.

• Air quality problems can be caused by the HVAC system maintenance procedures if these operations dislodge accumulated dirt.

2-10 System Commissioning

Some leakage testing and air-side balancing is required to certify that the HVAC system and the structural envelope will perform as designed, and a functional check should be made on the equipment and controls. It also is important to present the home-owner with a package of information regarding the equipment that has been installed and documentation of the tests that have been performed.

- Blower door tests can be used to evaluate the tightness of the structural envelope.

- Blower door or duct blaster tests can be used to evaluate the tightness of a duct system installed in an unconditioned space.

- The air flow associated with each supply outlet and return should be measured and, if necessary, adjusted.

- In some cases a fan speed adjustment will be associated with the air balancing work.

- Combustion efficiency tests are required to certify that fuel-burning equipment is operating safely and efficiently.

- Combustion air tests (as per the NFGC, Appendix H or Standard 62-1989, Appendix B) are required to certify that an adequate amount of air will be available for atmospheric burners.

- Operating and safety control cycles should be tested.

- Copies of the test data should be presented to the owner.

- Operation and maintenance procedures should be documented and presented to the owner.

- Copies of the design calculations should be presented to the owner.

- Copies of the equipment manufacturer's engineering information, installation instructions, and parts lists should be presented to the owner.

- A service contract should be offered to the owner.

2-11 Radon

Radon gas is an odorless, colorless, carcinogenic pollutant (as classified by the Environmental Protection Agency) that is produced by the natural decay of unstable elements (radium and uranium) that are contained in the soil, rocks, and water that surround a home. Since the concentration of these source elements varies from locality to locality, homes are selectively subjected to this type of contamination.

If radon is present in the surrounding soil, it can be drawn into a basement, crawl space, or another part of the home by pressure differences. This undesirable circumstance can be detected by testing.

- Short-term screening tests feature the use of inexpensive detectors and sampling periods that last from two to seven days. These tests are useful only for detecting the potential for a radon problem.

- If a screening test returns a radon reading of 4 Pico-curries per liter or more, a comprehensive evaluation will be required. This process could involve a series of short-term tests taken at different times and locations, over a period of a year, or a continuous long-term test.

If skillful testing verifies that a home has a radon problem, mitigation will be required. This work usually features "strategies designed" to keep the offending gas out of the home. This goal can be achieved by sealing the structure and by controlling the pressure differences across the structural components, particularly in the foundation areas.

- Seal dirt floors.

- Seal cracks in concrete floors and concrete walls.

- Seal cracks and pores in block and masonry walls.

- Trap sumps and floor drains.

- Install special well and sump-pump covers.

- Seal piping and electrical penetrations.

- Use a ventilation system to pressurize the structure.

- Use an exhaust system to create a negative pressure at the outside of foundation surfaces.

In cases where source control fails to resolve the problem, a mechanical or natural ventilation system may be required to control the concentration of the contaminate. Also note that some air cleaning devices have radon-capture capability.

In any case, radon problems concern comfort-conditioning contractors because of the interfaces associated with controlling space pressures and insuring acceptable indoor air quality. Since testing and mitigation require specialized knowledge, the contractor has the option of acquiring the necessary skills or working with an agent that specializes in this service.

Section 3
Zoning

In residential applications, a good number of "cold room, hot room" complaints are caused by the performance limitations of single-zone conditioning systems. No matter how carefully this type of system is designed and installed, space temperature excursions will occur when two or more rooms are subject to an incompatible set of load conditions. These conflicting load patterns may be caused by the architecture, the position of the sun (time of day), or the type of load (heating or cooling).

The information in this section can be used to decide if a single zone-system is an appropriate choice for a particular home. And, if a multi-zone system is required, this information can be used to identify the rooms that have similar load characteristics (so they can be grouped into zones).

3-1 Zoning Not Required

A single-zone control strategy is appropriate for single story homes that have a compact floor plan, providing that the rooms are "open" to each other. In this case, natural convection currents tend to minimize room-to-room temperature differences. For small rooms, an open door is adequate for establishing the required connectivity. For large rooms, an opening that is equal to or greater than 25 percent of the partition area is required.

3-2 Zones Created by Isolation

When rooms are isolated from each other (by a closed door or an inadequate opening), room-to-room temperature differences cannot be moderated by natural convection. In this case, the desired temperature can only be maintained in the spaces that are under direct control of the thermostat and in other rooms or spaces that happen to have a similar load condition.

3-3 Incompatible Load Patterns

Incompatible load patterns cause the thermostat to be satisfied when one or more rooms still require heating or cooling. This condition will occur whenever an isolated room has a load pattern that is not synchronized with the load that affects the thermostat (as the day progresses from morning to evening or as the season changes from cooling to heating).

Time-of-Day Load Patterns
Conflicting time-of-day load patterns are caused by the variation in solar gains as the sun moves across the sky. Temperature control problems should be anticipated if the thermostat is placed in a room that is subject to a large solar load, because the condition in that room are not usually representative of the conditions in all the rooms. Conversely, if the thermostat is placed in a room that is not subject to a large solar gain, temperature control problems can occur in the rooms that have significant solar gains. It follows that the most severe temperature control problem will occur when isolated rooms and the room that is under thermostatic control are affected by large solar gains that reach a peak during different hours of the day.

Seasonal Load Patterns
Conflicting seasonal load patterns are a result of envelope designs that cause some rooms to require relatively more winter heating than summer cooling, while other rooms require relatively more summer cooling than winter heating. For example, north facing rooms with a large glass area can have a relatively large heating load and a moderate cooling load, while west and south facing rooms with large glass areas can have a relatively large cooling load compared to the heating load. Large heating loads and small cooling loads also are typical of rooms that have large exposed surface areas and relatively small solar gains. For example a north-facing room that has a slab floor will have a larger heating requirement and smaller cooling requirement than a west-facing or south-facing room located over a basement.

Incompatible seasonal load patterns can be discovered by comparing the percentage of the blower CFM that is required to heat or cool a room. An example of this type of data is provided by Figure 3-1.

	Room Loads and Load Fractions						Entire House
	1	2	3	4	5	6	
Cooling	4.5	5.0	3.0	5.9	4.6	2.1	25.1
% Total	18	20	12	24	18	8	
Heating	6.3	7.0	6.5	6.2	7.0	2.8	35.8
% Total	18	20	18	17	20	8	
T - stat				Yes			
Load values represent 1,000 BTUH							

Figure 3-1

Reference to Figure 3-1 indicates that the thermostat is located in room-4 and that 24 percent of the blower CFM is required to neutralize the room-4 cooling load during a hot summer day. However, on a cold winter day, room-4 only requires 17 percent of the blower CFM. Therefore, if the air flow has been adjusted for cooling, room-4 will receive 7 percent more heating capacity than it needs, but room-4 will not overheat because the thermostat will compensate for the excess supply CFM by reducing the equipment run-time. However, the temperature in some of the other rooms may drift out of control — for example, consider room-3.

Room-3 requires 12 percent of the cooling capacity. On a summer design day, the temperature in this room will be maintained because room-3 and -4 both receive the correct portion of the blower CFM. However, in the winter, room-3 requires 18 percent of the blower CFM. This means that on a cold winter day — even with the equipment operating continuously — room-3 will be 6 percent short of heating capacity. Also note that the heating problem in room-3 will be acerbated by the behavior of the room-4 thermostat, which tends to reduce the equipment run-time.

Of course, the room-3 temperature problem could be resolved by moving the thermostat to room-3, but this strategy will create an overheating problem in room-4. (Since room-3 is 6 percent short of heating capacity, a room-3 thermostat will call for continuous equipment operation during a very cold day even though room-4 is receiving more supply air than it needs.)

The zoning problems associated with the seasonal air flow requirements are much less likely to occur when all the airflow adjustments move in the same direction. In this case, an increase or decrease in the equipment run-time will globally compensate for deficient or excessive airflow. (Since the seasonal airflow requirements are dictated by architectural features and construction details, it is possible to design a home that minimizes the seasonal changeover problem.)

3-4 Buoyancy Problems

In multi-story structures, warm-buoyant air will rise to the upper level, and cool-dense air will flow down the stairs or cascade over the balcony and stratify near the floor of the lower level. Since the thermostat that controls a single-zone system is usually located at the lower level, the upper level tends to overheat and the lower level tends to be too cool.

3-5 Architectural Review

The zoning problems that are created by incompatible time-of-day or seasonal load patterns and buoyancy effects can be anticipated by reviewing the architectural features of the home. This survey should consider the glass areas, number of levels, floor plan, location of the rooms, location and orientation of exposed surfaces, and thermal characteristics of structural components.

- Any room that is isolated from the space that is controlled by the thermostat is a candidate for zone control.

- Split-level and multi-level homes are always candidates for zone control.

- Rooms that have picture windows, skylights, and sliding glass doors normally require zone control when the glass area exceeds 30 percent of the wall area. Zoning is mandatory when a room features one or more walls that are mostly glass.

- When a home spreads out in many directions, room-to-room and room-to-house load ratios will continuously change throughout the day and night, and they will be affected as the season changes from summer to winter.

- Even symmetrically shaped structures can be subject to a considerable variety of load conditions and load ratios if the home is relatively large.

- Attic rooms and rooms located over a garage, patio, or car port are candidates for zone control because these rooms have a greater amount of exposed area than the rooms associated with the rest of the structure.

- Basement rooms have unique heating and cooling requirements because the heating and cooling loads are affected by the thermal mass of the earth.

- Rooms located over a heated basement are not compatible with rooms that feature slab construction or an exposed floor.

- A zone may be created when one part of the building envelope features insulation materials and construction techniques that are considerably more, or less, efficient than the rest of the envelope.

- Rooms containing a pool or hot tub should always be isolated from the rest of the structure and from the central heating and cooling system.

3-6 Occupancy Considerations

Zone control is required when the homeowner demands the same level of comfort in every room or the ability to control temperature on a room-by-room basis to the satisfy each family member. Zone control also is desirable when, on a regular basis, one or more rooms are used to entertain a large number of people.

3-7 Economics of Zone Control

Zoned systems cost more to install than a single-zone system, but they can be much more economical to operate. Energy savings could range from 10 percent to more than 30 percent. Maximum savings are realized when the controls maintain the occupied zones at the design temperature while the unoccupied zones are setup or setback.

3-8 Load Pattern Compatibility

Two or more rooms can be controlled by a single thermostat if the rooms have a similar time-of-day load pattern and a compatible season-to-season load pattern. As noted in Section 3-1, these conditions are normally satisfied when a group of rooms are open to each other. It also is possible for these conditions to apply to a suite of isolated rooms when load pattern compatibility can be demonstrated by the following logic and procedures.

Time-of-Day Load Patterns

Time-of-day load patterns mostly depend on solar gains. This means that glass area and compass direction are the primary factors that determine time-of-day compatibility. The following guidelines are based on the zone load adjustment factors published in the Appendix 2, **Manual J**, 7th edition. (This table indicates that, depending on glass area and direction, the peak time-of-day cooling load may exceed the traditional **Manual J** cooling load estimate by 10 to 35 percent.)

- If the glass area exceeds 30 percent of the wall area, rooms that are not coupled (by convection through adequate openings) to the space that is controlled by a thermostat should be treated as a separate zone.

- Isolated rooms that have glass areas that are less than 30 percent of the wall area can be grouped together if they have compatible exposures.

- Rooms that have small solar gains (due to a north-facing exposure, a small glass area, or some form of external shading) are not subject to solar load fluctuations, so they can be grouped together.

- Table 3-1 evaluates the time-of-day compatibility of two or more rooms. For example, the table indicates that a west-facing room is not compatible with an east-facing room or a north-facing room, but it is compatible with a south-west-facing room.

Seasonal Load Patterns

The seasonal load patterns of a group of rooms can be evaluated by dividing the room cooling load by the room heating load (C/H ratio). Rooms that have a C/H ratio that is within 15 percent of the room being controlled by the thermostat can be considered to have compatible seasonal load patterns.

Time-of-Day Compatibility								
	N	NE	E	SE	S	SW	W	NW
N	—	Yes	Yes	Yes	No	No	No	No
NE	Yes	—	Yes	Yes	No	No	No	No
E	Yes	Yes	—	Yes	No	No	No	No
SE	Yes	Yes	Yes	—	No	No	No	No
S	No	No	No	No	—	Yes	No	No
SW	No	No	No	No	Yes	—	Yes	No
W	No	No	No	No	No	Yes	—	Yes
NW	No	No	No	No	No	No	Yes	—

Rooms that have glazing areas that exceed the 30 percent of the wall area should be a separate zone.

If a corner room has an exposure with a large glass area and a small glass area, use the compass point that corresponds to the large glass area.

Use a common compass point for corner rooms that have about the same amount of glass area on each exposure. For example, a room with similar west and south glass areas should be treated as a south-west exposure.

Table 3-1

For example, Figure 3-2 summarizes the **Manual J** information associated with six isolated rooms. This figure also shows the corresponding C/H ratios and a set of values that are 15 percent higher and lower than the room-2 C/H ratio. These calculations indicate that if the thermostat is located in room-5, rooms-1 and -2 can be zoned with room-5.

	Room Loads and C / H Ratios					
	1	2	3	4	5	6
Cooling	4.5	5.0	3.0	5.9	4.6	2.1
Heating	6.3	7.0	6.5	6.2	7.0	2.7
C/H	0.71	0.71	0.46	0.95	0.65	0.78
85 %					0.55	
115 %					0.75	
Load values represent 1,000 BTUH						

Figure 3-2

There will still be some temperature excursion in the rooms that are not under direct control of the thermostat, even if they have C/H ratios that meet the 15 percent requirement. At the **Manual J** design conditions, a 15 percent difference in the C/H ratio can translate into a 2 to 4 degree room-to-room temperature difference.

The worst control point always will be associated with a room that has a relatively high or low C/H ratio, especially when there is a considerable spread in the C/H ratios. For example, Figures 3-3 and 3-4 show that if a thermostat is installed in room-3 or room-4, none of the other rooms will have a C/H ratio that is compatible with either of these control points.

	Room-3 Control Point					
	1	2	3	4	5	6
Cooling	4.5	5.0	3.0	5.9	4.6	2.1
Heating	6.3	7.0	6.5	6.2	7.0	2.7
C / H	0.71	0.71	0.46	0.95	0.65	0.78
85 %			0.39			
115 %			0.53			
Load values represent 1,000 BTUH						

Figure 3-3

	Room-4 Control Point					
	1	2	3	4	5	6
Cooling	4.5	5.0	3.0	5.9	4.6	2.1
Heating	6.3	7.0	6.5	6.2	7.0	2.7
C / H	0.71	0.71	0.46	0.95	0.65	0.78
85 %				0.81		
115 %				1.09		
Load values represent 1,000 BTUH						

Figure 3-4

Figure 3-5 demonstrates the benefit of a mid-range control point. Notice that if the thermostat is located in room-1 or room-2, rooms-5 and -6 will have C/H ratios that are compatible with either of these control points. And, even though rooms-3 and -4 are not compatible with these control points, the temperature swings in these rooms will be less severe because the equipment run-time will be determined by a control point associated with a mid-range C/H ratio.

	Room-1 or Room-2 Control Point					
	1	2	3	4	5	6
Cooling	4.5	5.0	3.0	5.9	4.6	2.1
Heating	6.3	7.0	6.5	6.2	7.0	2.7
C / H	0.71	0.71	0.46	0.95	0.65	0.78
85 %	0.60	0.60				
115 %	0.82	0.82				
Load values are in 1000s of BTU per hour						

Figure 3-5

The **Manual D** cooling and heating CFM values can be used to select the best location for the thermostat. This is done by calculating the difference between the two air delivery requirements (heating and cooling). The best location for the thermostat is associated with the rooms that have the smallest difference between the cooling CFM value and the heating CFM value.

3-9 Zoning Requirements Must Be Evaluated

A zoned system may be specified by a builder or homeowner. Or, a zoned system might be approved by a builder or owner if the contractor can demonstrate that zoning is required to meet a minimum performance standard, as far as comfort and energy costs are concerned. In these situations, the contractor obviously is obligated to evaluate the zoning requirements of the home because this analysis is part of the requested design work. However, many builders and owners are not this explicit.

It is more common for projects to have no explicit (written or verbal) performance specification. But, this does not mean that a builder or owner has no expectations regarding comfort. Therefore, the contractor should always evaluate the home's zoning requirements and educate the builder or owner about the consequences of installing a single-zone system. This preemptive analysis will ensure that no disappointment will be caused by unexpressed expectations that can not be satisfied by a single-zone system.

3-10 Procedure for Evaluating Zoning Requirements

Zoning is required when the loads monitored by a centrally located thermostat are not representative of the loads associated with isolated rooms. This situation will almost certainly occur if rooms, groups of rooms, or levels have incompatible architectural features, unsynchronized time-of-day load patterns, or conflicting seasonal load patterns. Zoning also is required to satisfy a request for temperature control on a room-by-room basis. The following procedure can be used to decide how a home should be zoned.

Step 1 — Evaluate Architectural Features
Zones are created when one or more rooms have incompatible architectural features. The following list summarizes what to look for.

- Floor plans that feature more than one level.
- Floor plans that spread out in many directions.
- Dwellings that only have one or two exposures.
- Rooms that have unusually large glass areas.
- Rooms in a basement.
- Rooms in an attic.

- Rooms above a garage or car port.
- Rooms over a ground slab
- Rooms over an exposed floor.
- Rooms that will be used for entertaining many people.
- Rooms that contain a hot tub or swimming pool.
- Spaces used for a solarium or atrium.

Step 2 - Evaluate Time-of-Day Load Patterns

Use the compatibility-by-exposure information (Table 3-1) to evaluate the effect of solar loads. Rooms that have similar time-of-day load patterns can be combined into the same zone.

Step 3 - Evaluate C/H Ratios

Calculate the C/H ratio for each room. Rooms with similar C/H ratios can be combined into a common zone. (The C/H ratio of a room that is associated with a zone must be within 15 percent of the C/H ratio of the room that will be equipped with the zone thermostat.)

Step 4 - Evaluate the Room-to-Room Openings

Small rooms can be combined into a common zone if they are connected by an open door. (Of course, a zone is created whenever the door is closed.) For large rooms, connectivity is established by an opening equal to or greater than 25 percent of the area of the common partition.

Step 5 - Finalize the Zoning Plan

Rooms located on different levels are not compatible, so zoning decisions must be made on a level-by-level basis. A one-zone-per-level strategy will effectively deal with the temperature control problems associated with the buoyancy of warm air, but it will not resolve the problems caused by incompatible time-of-day load patterns, seasonal load patterns, and unique architectural features. The following guidelines apply to rooms that are on the same level.

- Rooms that have exceptional features must be treated as separate zones. Examples are a room with a large amount of glass or a basement room.

- Rooms that are open to each other can always be combined into a common zone (even if the time-of-day and seasonal load patterns are not compatible) provided that there are no unusual features.

- Rooms isolated from each other can be combined into the same zone, provided that they have compatible time-of-day load patterns and similar C/H ratios (and no unusual features).

3-11 Multi-Level Homes

Multi-level construction creates incompatible load patterns in the vertical direction. For example, the upper level of split-level homes tend to be too hot, the middle level reasonably comfortable, and the lower level too cold. This tendency also occurs in homes that have two or four levels. Obviously, the problem is associated with the buoyancy of warm air, open stair wells, balconies overlooking a vaulted room, and the location of the thermostat. Other contributing factors might include a slab floor or entrance at the lower level and an exposed roof at the upper level. Since there could be two to four separate areas that have distinctively different comfort conditioning requirements, these type of homes are candidates for level-by-level zone control, with a thermostat located on each floor.

In some cases the number of zones may be greater than the number of levels because architectural variety generates additional zoning problems. For example, there are load pattern incompatibilities between the rooms in the main body of the home and basement rooms; rooms over a garage; attic rooms; and rooms located over a patio, carport, or crawl space. In addition, some rooms may be enclosed by framed walls and some may feature a masonry boundary.

Large glass areas also must be considered. Most homes have at least one large picture window and a sliding glass door, but elaborate architecture may flaunt multiple examples of this type of construction. And, when the glazing treatment is taken to the limit, one or more rooms may have an entire outside wall made of glass. In any case, when rooms with large window areas are exposed to the sun, they can overheat (if the thermostat is in another area of the home) or cause other rooms to be too cold (if the thermostat responds to the solar heating effect). Also note that during cold nights, these interactions will be completely reversed as large glazing areas become a major source of heat loss.

3-12 Rambling Floor Plans

Some one-level homes spread out in all directions. This means that the various wings and projections will likely have significantly different reactions to the sun, wind, and outdoor temperature. Therefore the room-to-house load ratios will not be the same throughout the day and night, or from summer to winter. These houses cannot be properly conditioned by a single-zone system because the structure is equivalent to two, three, or four separate homes.

3-13 Town Houses

Town houses are difficult to condition because of the gravity-driven air exchange that occurs among the different levels. And, these floor-to-floor temperature difference problems are acerbated when the floor of the lower level is below grade. Since there are at least three to four separate areas that have distinctively different comfort conditioning requirements, these dwellings are candidates for level-by-level zone control, with a thermostat located on each floor.

- Even if it is not used as a living space, the below-grade area should be completely insulated and heated because the temperature in this area affects the floor temperature on the next level. And if a below-grade area is used as a living space, maximum comfort will be provided by supply outlets located low and at the outside wall, complimented by a return at the bottom of the stairwell. Also consider using baseboard heat to supplement an overhead supply system.

- Try to control the air interchange between levels by installing effective returns in the stairwell landings. Also consider installing a ceiling fan in the stairwell.

- Buoyancy effects are reduced when level-to-level temperature differences are minimized. Use heating equipment that operates with a small temperature rise, make sure that all supply outlets are properly located and correctly sized, provide returns for each level and isolated areas, make sure that the air flows are balanced, and consider operating the blower continuously.

3-14 Condominiums

Condominium living spaces could be consolidated on one level or distributed across two to four levels. If multiple levels are involved, the comments made about multi-level homes and townhouses apply to this type of construction. These dwellings also are characterized by a limited number of exposures, possibly just one. Regardless of the number of levels, an interior-perimeter zoning strategy is appropriate when a perimeter area is subject to a large solar gain.

3-15 Get the Most Out of Single-Zone Designs

Zoned systems provide close control of temperature on a room-by-room basis, but they are more expensive than single-zone systems. If limited capital precludes zone control, careful design work will maximize the level of comfort that can be provided by a single-zone system.

Make Accurate Load Calculations
Base the load calculations on the recommended **Manual J** design temperatures (Table 1) and take full credit for insulation levels and efforts to seal the envelope. Take credit for internal shading (blinds, drapes, or roller shades) and external shading (shade screens, awnings, and overhangs).

Design For Low Humidity
Base the **Manual J** cooling latent load calculation on the "grains difference" value associated with the 50 percent indoor humidity value (Table 1).

Use Moderate Supply Air Temperatures
Minimize the buoyancy problems that cause floor-to-floor and ceiling-to-floor temperature differences and maximize the performance of the supply air outlets (as defined by throw and drop) by using heating equipment that operates with a small temperature rise.

Do Not Oversize Equipment
Use manufacturer's performance data to obtain the closest possible match between equipment capacity (sensible, latent, and heating) and the calculated loads.

Use an Appropriate Distribution System
Perimeter air distribution systems are preferred for cold-climate heating and ceiling outlets are best for warm-climate cooling. Refer to **Manual T** for comprehensive guidance on this subject.

Calculate the Duct Sizes
Use the **Manual D** duct sizing procedure to ensure that an adequate flow of air is delivered to each room.

Minimize Duct Losses
Insulate and seal duct runs and equipment cabinets that are located in an unconditioned space.

Select the Supply Outlets Carefully
Install supply air outlets that are compatible with the construction and the climate, and make sure that they will thoroughly mix the conditioned supply air with the room air. (As explained in **Manual T**, supply air outlet performance depends on location, size, and the type of device.)

Provide an Adequate Return System
A low-resistance return path is required for every room or space that receives supply air. (Central returns can be used for spaces that are open to each other. Individual returns or transfer grilles are required for rooms that can be isolated by closing an interior door. If more than one level is involved, use at least one return for each level.)

Balance the System for Seasonal Conditions
If there is a wide spread in the C/H ratios, the supply-air flow rates should be readjusted when the season changes from cooling to heating and vise versa. Or, if it is too inconvenient to make two damper adjustments per year, the air balance work should be based on room airflow rates that fall midway between the heating CFM and cooling CFM values generated by the **Manual D** procedure.

3-16 Continuous Blower

For comfort's sake, continuous blower operation can noticeably improve the performance of a carefully designed single-zone system. This tactic will reduce the room-to-room and level-to-level temperature differences caused by incompatible time-of-day load patterns, seasonal load patterns, and buoyancy effects because the indoor air will be mechanically blended on a continuous basis. But, there are some drawbacks associated with operating the blower continuously.

- If the various load excursions are severely out of balance, there is no guarantee that room-to-room or level-to-level temperature differences will be adequately controlled.

- Operating costs escalate (maybe by $50 to $100 per year) when the fan operating hours increase.

- If the coil and drip pan are wet when the system operates in the cooling mode, moisture will be added to the recirculated air when the compressor cycles off. In some cases, this mode of operation can cause a high humidity level in the conditioned space. (Re-evaporated moisture will not be a problem if the humidity is relatively low when the compressor is operating. So, design for 50 percent relative humidity and do not oversize the cooling equipment.)

3-17 Supplemental Equipment

For comfort, single-zone system performance can be improved by installing supplemental devices in problem areas. (In this case, a zoned system is created by adding self-contained, locally controlled equipment to a central system.)

For example, electric baseboard radiation can be used to maintain comfort in a room that requires relatively more heat than the space that has the central thermostat. Or, a small split system (or window unit) might be used locally to supplement the cooling provided by a central air handler. Or, a volume damper (and control) could be used to divert cooling capacity from the unoccupied rooms to occupied rooms that are subject to temperature excursions.

3-18 Additional Information about Zoned Systems

Additional information about zone control for residential structures can be found in **Manual J** and **Manual D**. Refer to Appendix 2 and Appendix 3 in **Manual J** for applicable load calculation procedures; and to **Manual D**, Section 1 for information about why zoning is required, and Section 11 for information about duct sizing procedures that apply to variable-volume systems.

Section 4
Thermal Envelope

Efficient building practices are the most cost-effective way to maximize comfort, control HVAC system installation costs, and minimize HVAC system operating costs. In this regard, the construction quality is defined by the ability of the thermal envelope to regulate heat flows, air leakage, and moisture migration.

4-1 Comfort Issues

Even if a home is drafty and poorly insulated, the HVAC system can be designed to maintain a certain temperature at the thermostat, but this does not mean that the occupants will be comfortable. As explained in Section 1, discomfort is caused by drafts and cold walls, glass surfaces, and floors. Therefore, comfort is as closely linked to construction quality as it is to the performance of the HVAC system.

4-2 Zoning Problem

If the home is conditioned by a single-zone system, the room-to-room temperature differences tend to be unacceptably large if the load patterns associated with satellite rooms are not compatible with the load pattern acting on the room with the thermostat. These load pattern incompatibilities are acerbated when the thermal envelope is leaky and poorly insulated. (See Section 3 for a discussion of zoning issues.)

4-3 Economic Issues

There is an inverse relationship between construction cost and operating cost, and there is a similar link between construction costs and comfort. Therefore, home builders and homeowners always are required to compare the marginal increase in construction cost with the benefit delivered by a more efficient envelope. Of course, these decisions also depend on the project manager's short-term and long-term goals.

- If low construction costs are the primary goal, the efficiency of both the structural envelope and the HVAC system will be dictated by building codes, utility regulations, and minimal industry standards.

- If a low operating cost is desired, the marginal benefits associated with building a more efficient structure and using a more efficient HVAC system must be weighed against the incremental increase in the installation cost.

- If comfort is the issue, the incremental increase in the cost of building a more efficient structure and using a more

sophisticated HVAC system must be weighed against the anticipated increase in comfort.

Notice that regardless of the mission, it is always necessary to ask whether it is more cost effective to enhance the thermal performance of the building or improve the capabilities of the HVAC system. Of course, the answer to this question is conditional because it depends on the performance levels established by codes and regulations, type of climate, construction costs, equipment costs (installed), maintenance costs, utility rates, utility incentive programs, financing costs, and tax policy. However, it is possible to make some general statements.

- At a particular site, the amount of energy (BTU per year delivered or extracted from the conditioned space) required to heat and cool a home is determined solely by the thermal envelope efficiency. Therefore, as far as comfort and operating costs are concerned, it is always good practice to make a substantial investment in an efficient structural package.

- An effective level of envelope efficiency can be obtained by providing a reasonable amount of insulation, installing competitively priced window and door assemblies, and using standard sealing techniques. If more efficiency is desired, the additional costs of installing state-of-the-art components and using innovative techniques should be justified by calculations that indicate an acceptable return-on-investment.

- The amount of energy that must be purchased from a utility depends on the amount of energy required to heat and cool the home, the efficiency rating (SEER, HSPF, COP, or AFUE) of the HVAC equipment, and the efficiency of the distribution system. In this regard, it is always cost-effective to insulate and seal a duct run or equipment cabinet located in an unconditioned space, and this also is important for comfort. However, the use of state-of-the-art equipment and unconventional system concepts should be justified by a return-on-investment calculation. In this regard, the cost-benefit calculations may be biased by a utility incentive program that encourages designs that otherwise would not be cost effective.

4-4 Insulation

Everyone understands that a properly insulated home will be comfortable and economical to operate, but there are always questions about product options, product performance, and the optimum amount of insulation. The answers to these queries depend on the insulation's location, the performance characteristics of the insulation material, and the relationship between the installed cost and the annual fuel cost.

Traditional Insulation Products

As indicated by Figure 4-1, the most common types of insulation are batt or blanket, loose-fill, blown fibers or particles, various types of rigid sheets, and chemical foams. All of these products are designed to reduce the heat flow through a structural assembly, and except for some rigid sheets, these products do not provide structural support.

Reflective Foils and Coatings

Reflective coatings reduce the amount of radiant heat transfer. They often are combined with a board or sheet material that is used to reduce the conductive heat flow. It is important to understand that the reflective surface must face an air space that is at least three-fourths of an inch wide in order to be effective.

Radiant Barrier

A radiant barrier is a non-traditional insulation product that is normally installed in homes located in hot sunny climates. It consists of a panel coated with a bright foil backing. When installed, it becomes part of the roof assembly (with the foil-side facing down).

Radiant barrier technology is different because it reduces the radiant heat transfer between the inside of the roof surface and the top of the attic insulation. This, in turn, will reduce the cooling load associated with the ceiling panel that is below the attic insulation. The barrier also will reduce the cooling season duct losses (due to wall conduction and return-side leakage) because the attic temperature will be somewhat lower during hot sunny days (compared to an attic that is not equipped with a barrier).

It is important to understand that as far as cooling loads and attic temperatures are concerned, radiant barriers are only effective when the sun is heating the roof to temperatures that are significantly warmer than the outdoor air temperature.

There is no benefit when there is no solar effect — late at night, for example.

Also note that a radiant barrier has no effect on the ceiling heat flows (cooling or heating) produced by the temperature difference between the attic and the conditioned space. Therefore, ceiling insulation still is important because these temperature differences will be at least as large if the attic is properly ventilated as the outdoor-indoor temperature difference (or larger when the sun heats the roof during the summer).

Aerated Concrete Blocks

Aerated concrete blocks provide a second example of an innovative insulation product. These blocks are unique because a structural material provides an insulating benefit. This is accomplished by saturating a block of concrete with little air pockets. The result is a lightweight product that has the strength required for a load-bearing wall and a thermal resistance that is significantly larger than the resistance of other types of masonry materials.

Rating Insulation Materials

Insulating product ratings are based on laboratory tests. These tests evaluate a product's thermal resistance by measuring the heat flow produced by a constant temperature difference. The results of these tests are published as the R-value rating. This rating may apply to just one inch of material or it may refer to the total thickness of a product. For example, fiberglass batt material has a per-inch rating of about R-3.14, and some of the batts designed to be compatible with a 2" x 4" stud space have a rating of R-13 or higher.

Note that an R-value is an absolute rating that applies to a steady-state condition, which means that during the test, the temperatures on either side of the material never change with time. In this respect, any claim of an "equivalent R-value" is subject to skepticism because it is based on a time-dependent heat flow pattern. In other words, legitimate R-values can be generated only by performing the time-invariant tests that are specified in the ASHRAE and ASTM standards.

Traditional Insulation Products				
Material	**Installation**	**Material**	**Vapor Barrier**	**Structural Benefit**
Batt or blanket	Piece work	Fiberglass, mineral wool	May be attached or separate membrane	None
Loose fill	Blown or poured	Fiberglass, rock wool, vermiculite, perlite, celluose	Separate membrane	None
Rigid board	Piece work	Fiberglass, polystyrene, polyurethane, plastic wool, vermiculite, perlite, organic cellulose	May be attached or separate membrane	Sheathing
Chemical foams	Sprayed	Synthetic	Closed cell desirable	May add stiffness

Figure 4-1

The sophistry associated with an equivalent R-value results from a desire to take credit for the thermal mass of the structural material. (There are valid reasons for evaluating the effect of thermal mass, but they are not related to the steady state performance of an insulating material.)

The thermal mass of a material depends on its specific weight and specific heat. The benefit of thermal mass is realized when one side of the insulating material is subjected to a temperature that changes with time, especially if the temperature-time relationship is periodic (a 24-hour sine wave, for example). In this case, if two materials have the same R-value, there will be less heat flow associated with the material that has the larger thermal mass.

In the **Manual J** cooling load calculation procedure, the thermal mass effect is modeled by using an effective temperature difference that is smaller than the actual air-to-air temperature difference across the structural component (refer to the partition ETD values listed in Section 7 of **Manual J**). However, thermal mass is not relevant to the **Manual J** heat loss estimate because this calculation is based on a constant temperature difference model that represents a worst-case scenario (short winter days with negligible solar loads).

Also note that different philosophies are associated with the calculations that are used to size heating and cooling equipment and the calculations that estimate energy consumption. In the first case, simple static models are acceptable because equipment size is based on a single, steady-state load condition. In the second case, dynamic conduction models are required because energy estimates require an integration of the instantaneous load-temperature relationship over time.

Installed R-Value
When an insulating material is installed, the in-service R-value may be less than the laboratory-measured R-value. This difference occurs when the installed configuration is not identical to the tested configuration. For example:

• The R-value decreases when batt material is compressed.
• Infiltration through batt insulation decreases the R-value.
• Infiltration barriers preserve the R-value of batt insulation.
• The R-value of blown fill depends on the density of the fill.
• Moisture retention can affect the R-value of the insulation.
• Vapor retarders can improve the installed R-value.

Cost Issues
The costs of installing insulating materials escalate in some relation to the increase in the R-value. Therefore, once code requirements are satisfied and comfort issues are resolved, justification for an additional increment of insulation should

be based on the anticipated reduction in the annual energy costs.

There is no doubt that radiant barriers have the ability to reduce the peak cooling load on sunny days. And, it has been demonstrated that this reduction can lower operating costs if the home is located in a warm sunny climate. But, cost-benefit questions are difficult to answer because the effectiveness of the radiant barrier depends on the local climate and the R-value of the ceiling insulation. (The radiant barrier's effectiveness increases when the cooling costs are significantly higher than the heating costs, and decreases as the R-value of the ceiling insulation increases.) Also note that the solution to this return-on-investment problem is complicated by an attic duct system. (When attic duct runs are involved, the cooling-mode benefit of cooler attic must be balanced against the heating-mode penalty associated with a colder attic.) Ultimately, the designer must decide whether it is more cost effective, on an annual basis, to install a radiant barrier or to use more ceiling insulation, more duct insulation, and more duct sealant material.

Research indicates that clean radiant barriers can reduce the mid-summer ceiling load by 20 to 40 percent, depending on the amount of attic insulation. Since mid-summer ceiling loads typically represent 10 to 20 percent of the total cooling load, this reduction will translate into a 2 to 8 percent reduction in the total design-day cooling load. This does not mean that there will be an equivalent reduction in the annual operating cost, because a home is exposed to a wide variety of weather conditions during the year. For example, if 50 percent of the energy budget is spent on cooling, the corresponding reduction in cooling costs will range from about 1 percent to about 4 percent.

Test also indicate that attic temperatures are 10 °F to 20 °F lower when a radiant barrier is installed. Obviously, this will affect the efficiency of the duct system (particularly if it is poorly insulated, leaky, and located in an attic subjected to a lot of hot sunny weather), but the net reduction in the annual energy budget is extremely difficult to quantify, even for a specific home.

Selecting Insulation Materials
There are many different insulation products, and builders must decide which product is the most suitable for a particular application. Figure 4-2 (on the next page) provides a list of the properties that affect the selection of insulation materials.

Code Requirements
Insulation requirements depend on the local climate. These requirements pertain to materials, points of application, insulation levels, and installation procedures. Normally, building codes and utility regulations provide climate-spe-

Insulating Product Performance Issues		
Property	**Issue**	**Desired Tendency**
Resistance per inch	Amount of space required for installation	Large
Weight per unit volume	Structural loads	Small
Thermal mass	Heat storage ability moderates load swings	High
Reflective coating	Reduced radiation across an air space	Bright Aluminum Foil
Flame spread rating	Fire hazard	Low
Volatility (off-gasing)	Indoor air quality	Low
Compressibility	Affect on R-value	Low
Permeability	Stop moisture migration	Low
Absorptivity	Affects R-value, mold, mildew, rot, air quality	Low
Dimensional stability	Wrapping, seam separation, splits and cracks	High
Durability	Damage by people and animals	High
Workability	Ease of installation	High
Installed cost	Return on investment	Low

Figure 4-2

cific guidance. Also note that a code or regulation could be based on some type of cost-benefit analysis, or it might represent the consensus opinion of a group of building industry stakeholders.

For the amount of insulation, the code may refer to the tested R-value of the insulating material, the overall R-value of the structural assembly that encloses the insulation material, or the overall U-value of the structural assembly. (The U-value is the reciprocal of the overall R-value.)

Insulating Techniques
Insulation techniques related to the type of product and the size of the R-value depend on the nature and location of the structural component. The following paragraphs provide a brief summary of insulating practices and code requirements on a component-by-component basis.

Grade Slabs
Most of the heat loss associated with a grade slab occurs at the exposed perimeter. These losses can be controlled by installing rigid board insulation that extends straight down, parallel to the slab edge; or by installing rigid board insulation on the face of the slab edge and then turning the board under the slab. In this regard, the code requirements usually pertain to the tested R-value of the board material and the buried depth, or the length of the horizontal projection under the slab. (Insulation below the entire slab is not necessary.)

Foundation Walls
Foundation walls include crawl space walls and basement walls that are entirely above grade, partially above grade, or entirely below grade. Insulation may be applied to the inside or the outside surfaces of these walls (or on both sides), or insulation may not be required if the wall encloses an unconditioned space.

- Insulation applied to foundation walls associated with open (or vented) crawl spaces would not be effective. When this type of construction is encountered, the insulation should be installed under the floor that covers the crawl space.

- Walls associated with unconditioned basements and enclosed crawl spaces do not have to be insulated, but then insulation may be required under the floor that covers the unconditioned space.

If insulation is applied to a foundation wall, it can be installed on the inside or the outside of the wall (or on both sides). Rigid board is compatible with either location and framing with batt insulation often is installed on the inside of foundation walls. The code requirements for this type of construction might specify the R-value of the insulation material, the composite R-value (or the U-value equivalent) of the structural sandwich, and when below grade surfaces are involved, the installed depth.

Block Walls
Rigid sheets of insulation can be applied to the outside or inside of a block wall (or on both sides), or if the wall is "famed-out," batt insulation can be added to the inside surface. The code requirement for above-grade block walls is usually related to the composite R-value (or the U-value equivalent) of the structural sandwich.

The R-values published for rigid foamboard represent the thermal performance of dry material. However, the R-value of certain types of rigid foamboard products will be reduced significantly if the material gets wet. This loss of performance is an important consideration when rigid foamboard is installed at the edge of a slab floor, or on the exterior surface of below-grade walls.

Rigid foamboard materials include extruded polystyrene, molded polystyrene, polyurethane, and foil-backed polyurethane. Extruded polystyrene is the only material that has a moisture-resistant closed cell structure. The other materials will absorb significant amounts of water, and as noted above, the insulating value will be degraded. The R-value is the most obvious performance parameter affected by water, but the time required for the material to dry out also is important. Figure 4-3 shows that extruded polystyrene is the preferred material for a wet environment.

Performance of Wet Foamboard		
Product	Percent of Dry R-Value	Below Grade Dryout Time
Extruded Polystyrene	99%	4 Days
Molded Polystyrene	73%	30 Days
Polyurethane	52%	100 Days

Figure 4-3

Framed Walls
Batt insulation usually is installed in framed walls, but blown fill also is used — particularly when insulation is added to an older home that has no wall insulation. In addition to the wall cavity insulation, rigid sheets of insulation can be applied to the outside or inside of the framing. The code requirement for framed walls usually is related to the composite R-value (or the U-value equivalent) of the structural sandwich.

Ceilings Below an Attic
Batt insulation typically is installed above exposed ceilings, but poured and blown materials also are common. Usually there are no rigid board products associated with a ceiling that is located below an attic. Regardless of the structural detail, the code requirement for ceilings below an attic normally pertain to the composite R-value (or the U-value equivalent) of the ceiling assembly.

Radiant barriers do effect the cooling-season heat gain associated with a ceiling below an attic, but they are not mentioned in most codes (as of 1996). If a code does address this subject, it could offer two ceiling R-value options — one for an attic without a radiant barrier and one for an attic with a radiant barrier.

Roof-Ceiling Assemblies
When a roof-ceiling assembly is insulated, blankets or batts typically are installed between the ceiling joists, but these cavities also can be filled with poured or blown materials. If rigid sheet insulation is installed (as the primary insulation or as a supplement to the cavity insulation), it might be installed on the top or the bottom of the ceiling framing. The code requirement for this type of ceiling construction usually is related to the composite R-value (or the U-value equivalent) of the assembly.

Roof on Beams or Rafters
Rigid board insulation normally is installed above ceilings that cover beams or rafters. In this case the finished surface (usually wood boards), sheathing, insulation and roofing materials create a compact, waterproof insulating package. The code requirement for this type of ceiling construction usually is related to the composite R-value (or the U-value equivalent) of the roof sandwich.

Floor Over an Unconditioned Space
When a floor is insulated, blankets or batts typically are installed between the floor joists. If rigid sheet insulation is installed (as a supplement to the cavity insulation) it normally would be found below the floor framing. The code requirement for this type of floor construction usually is related to the composite R-value (or the U-value equivalent) of the entire floor assembly.

Installation Issues
The performance of any insulating material is related directly to the skill and workmanship of the installer. The pertinent methods and procedures are documented in building codes, utility regulations, manufacturer's literature, and material published by insulation manufacturer's associations or home builder's associations. These references indicate that poor performance is associated with:

- Missing insulation
- Insulation that is squeezed or compressed
- Gaps, cuts, and tears in insulating material
- Gaps, cuts, and tears in the vapor barrier
- Poor fit around windows and doors
- Poor fit around electrical outlets and recessed fixtures
- Poor fit around exhaust fans and envelope penetrations
- Poor fit around supply air outlets and returns
- Materials subjected to moisture and wetting
- Wind-driven air flow through the insulating material
- Ineffective fasteners, staples, tapes, and sealants
- Inadequate attic ventilation
- Attic vents that are blocked by insulation material.

4-5 Vapor Barriers

If the air on one side of a structural component is more humid than the air on the other side, water vapor will flow from the humid side to the dry side. During cold weather, this flow will move in an indoor-to-outdoor direction because the moisture is released within the home (by the occupants, occupants' activities, humidifiers, and plants) causes the indoor air to be more humid than the outdoor air. During hot, humid weather, this flow usually moves in the opposite direction (out-to-in) if the home is air conditioned (except when the home is located in a very dry climate). It also is possible for water vapor to flow into the conditioned space from a crawl space or unconditioned basement, from the soil below a concrete slab, or from the ground next to a foundation wall. All of these moisture migrations are undesirable because they create comfort problems, structural problems, and air quality problems.

- During the cooling season, migrating moisture increases the latent load on the cooling equipment.

- Excessive moisture can produce a cold, clammy indoor environment.

- Moisture will condense within a structural sandwich if the temperature at some point within the sandwich is colder than the dewpoint of the air that is in the sandwich.

- Condensation within the structural sandwich can cause mold, mildew, fungus, and odor problems.

- Air quality problems are caused by biological growth within the structural sandwich and by moisture that migrates through foundation walls, crawl space floors, and slab floors.

- Condensation within the structural framing-insulation sandwich can rot or delaminate wood products, and damage insulating materials.

- Condensation within the structural sandwich can reduce the R-value of some insulating materials.

- Framing connections and masonry joints can be seriously damaged when the condensation within the structural sandwich freezes.

- Moisture can corrode metal parts of electrical systems and exhaust systems.

- Moisture migration can cause stains or damage the finish of exterior or interior surfaces.

- Moisture migration can cause paint to peel or it can damage exterior siding.

The moisture migration problems associated with heated, above-grade spaces can be controlled by installing a vapor retarder on the indoor side of the wall, ceiling, and floor insulation. (During cooling, in a hot coastal climate, migration can be controlled by locating the retarder at the outside surface of the insulation.) It also is necessary to place an impermeable membrane on the ground-side of slab floors.

Vapor retarders also are recommended for crawl spaces and basements. For crawl spaces, the walls should be sealed and an impermeable membrane should be installed over dirt floors and covered by a layer of limestone. (A concrete floor poured on top of an impermeable membrane is preferable to limestone). Basement moisture problems can be controlled by installing an impermeable membrane under the floor and coating the walls with a waterproofing material (preferably on the back-filled surface). In addition, on-grade and below-grade floors should be installed at a reasonable distance above the water table, and the site should be well drained and properly graded; all organic material should be removed during the site preparation work; all footers should be equipped with drains and covered with gravel back-fill; the top of the footers should protrude at least 8 inches above grade; and all wall and floor penetrations must be thoroughly sealed.

A vapor retarder may consist of an independent membrane installed over an insulated framework or a facing that is an integral part of the insulation product. In either case, there must be no cuts, rips, tears, or open seams in the membrane. Refer to building codes, utility regulations, and the literature published by the building industry and trade organizations for information about perm-ratings, material properties, and installation procedures.

Vapor retarders and impermeable barriers are just one aspect of a comprehensive moisture control strategy. Exhaust fans, vents, cooling coils, mechanical ventilation, and infiltration also have a significant effect on the amount of moisture contained in a conditioned space, crawl space, unconditioned basement, or attic.

- Exhaust fans (kitchen and bathroom) and a laundry room vent should be used to capture and expel moisture at the point of generation.

- Vents can be used to control the moisture that migrates into a crawl space (cold weather or dry climate).

- Vents must be used to control the temperature and humidity in attic spaces.

- Infiltration and ventilation reduces the indoor humidity during cold weather.

- Humidification equipment can be used to control the indoor humidity during cold weather.

- Infiltration and ventilation will increase the indoor humidity when the moisture content of the outdoor air is greater than the moisture content of the indoor air.

- Properly sized cooling equipment will keep the indoor humidity under control during the cooling season.

4-6 Windows and Doors

For windows and doors, there are two types of heating loads (conduction and infiltration), three types of sensible cooling loads (conduction, infiltration, and solar), and a latent cooling load (infiltration). In order to minimize the conduction and infiltration loads (and maximize comfort), windows and doors should have a low U-value (for the glass-frame assembly) and a low-test stand infiltration rate. It also is necessary to control solar gains. This can be accomplished by using special types of glass, surface coatings, internal shading devices (drapes, roller shades, or blinds), external shading devices (sun screens), overhangs, awnings, and porches.

Window Assembly U-Values
Window performance, as related to the overall U-value, depends on the number of panes in the assembly (one, two, or three), the gas that fills the space between the panes (air, argon, krypton, carbon dioxide, sulfur hexaflouride, or some mixture of these gasses), the coating applied to one or more of the glass surfaces (no coating or low emittance coating), the frame material (wood, aluminum, steel, fiberglass, vinyl, or some type of composite material), the continuity of the conduction path through the frame (thermal break or no thermal break), the architectural style (no muntins, external muntins or muntins between lights) and the ratio of the glass area to the casement area. The certified U-value of any window assembly currently being manufactured can be found in the "Certified Products Directory" published by the National Fenestration Ratings Council (NFRC). This directory also provides a detailed description of the construction features of each window assembly. (Window manufacturers also publish overall U-values for skylight assemblies.)

> Tight fitting drapes, blinds, and shades can reduce the U-value of a window assembly. However, credit is not taken for this effect because the insulating benefit is undocumented, and because there is no benefit when the device is open or loosely fitted to the frame.

Door U-Values
The U-value of a door depends on the size and performance of the glazing assembly that is installed in the door, the panel and framing materials (wood, aluminum, steel, fiberglass, or vinyl), and the insulating material between the door panels (air, solid panel, or some type of insulating material). The effective U-value of any door assembly currently being manufactured can be found in the manufacturer's product literature. (Certified U-values are obtained by performing the laboratory tests specified by manufacturer's associations.)

Crack Leakage
The "tightness" of a window or door assembly is evaluated by performing the laboratory tests required by window and door manufacturer's associations. Assemblies that have a leakage rate of 0.50 CFM per foot of crack length, when they are subjected to a 25 MPH wind, satisfy the industry standard for tightness, but upscale products have tested leakage rates considerably lower than this benchmark. When needed, the leakage rate of a particular product can be found in manufacturer's literature.

> On a per-square-foot basis, the conduction and leakage losses (or gains) of the most efficient windows, skylights, and doors are significantly larger than the losses and gains of a properly insulated and sealed wall or ceiling. Therefore the aesthetic benefits provided by windows and skylights always have to be balanced against the unfavorable effect that they have on the thermal efficiency of the structural envelope. Some codes and regulations address this issue, but if there is no official guidance, the use of glass will be dictated by the desires and economic resources of the owner or builder.

Shading Properties
The shading coefficient (SC) of window and skylight assemblies are calculated by dividing the solar gain of the candidate glazing assembly by the solar gain of a single pane of clear glass. These shading coefficients depend on the type of glass (clear, tinted, heat absorbing, or reflective), the number of panes of glass, and the type of coating (none, solar film, or low-emittance). Shading coefficient values can be found in window manufacturer's product literature.

> The shading coefficient is not the only parameter that affects window performance. The cooling load also depends on the glass area, the direction that the window faces, the internal shading device, and the external shading device. This means that architectural features and decorating decisions are as important, or possibly more important, than the shading coefficient of the glazing material. Some codes and regulations specify a minimum overall shading coefficient, based on the direction that the window faces, or limit the amount of glass area to some percentage of the floor area. If codes and regulations do not apply, the responsibility for producing an efficient glazing plan is transferred to the owner or builder.

> The primary function of a low emittance (Low-E) coating is to reduce the conductive heat loss associated with glass surfaces. As far as the cooling load is concerned, this coating produces a small reduction in the shading coefficient (SC) of the window assembly (when compared to a similar assembly that does not have a Low-E coating).

Cost Issues

Codes and regulations establish performance standards for window and door assemblies that are compatible with mainline products. If the glazing plan is held to a higher standard, the costs of installing state-of-the-art window assemblies should be balanced against the estimated reduction in the annual energy costs.

4-7 Crack Sealing and Air Infiltration Barriers

Heating, sensible cooling, and latent cooling loads are associated with an outdoor-indoor air exchange (infiltration). This leakage can occur at any unsealed crack or penetration in the thermal envelope. The driving forces that causes this leakage include the wind; buoyancy forces within the home; buoyancy forces associated with chimneys, vents and flues; combustion air drawn from the conditioned space by fireplaces and heating equipment; air expelled by kitchen and bathroom exhaust fans; air expelled through appliance vents; and the pressure conditions created when an air distribution system blower cycles on and off. A list of the most common leakage points and some rough estimates about the contribution to the total infiltration rate is provided:

- Exterior and interior walls (18% to 50%)
- Ceilings (3% to 30%)
- Windows and doors (6% to 22%)
- Fireplaces (0% to 30%)
- Exhaust fans, vents, and pipe penetrations (2% to 12%)
- HVAC duct systems and vents (0% to 40%)

Structural infiltration can be controlled by sealing the cracks and penetrations associated with the building envelope. This can be done locally by applying weather stripping, caulking, tapes, mastics, and sealants, or by wrapping the exterior envelope with an infiltration barrier. (A properly installed vapor barrier also will reduce the infiltration associated with exposed walls and ceilings.) Duct runs and air handler cabinets located in unconditioned spaces also must be sealed with suitable tapes and mastics, and exhaust vents and fireplaces should be equipped with back draft dampers. Also note that direct-vent heating equipment reduces infiltration because it does not draw combustion air from the conditioned space.

Building codes encourage tight construction because minimizing infiltration is easily justified by the corresponding reduction in energy use and operating costs. However, codes also emphasize that outdoor air is required to dilute pollutants and to control humidity during cold weather. This apparent contradiction does not mean that sealing efforts will be limited by the indoor air quality requirements, but it does mean that a natural or mechanical ventilation system will be necessary when infiltration cannot satisfy the code requirement for fresh air. (Some codes are based on ASHRAE Standard 62-1989. This standard recommends an infiltration rate equivalent to 0.35 air changes per hour or 15 CFM per person, depending on which criterion produces the largest flow rate.)

If a home is in the planning stage, the infiltration rate can be estimated by using the procedures documented in **Manual J**. These procedures also can be applied to an existing home, or the tightness of an existing home can be tested by using the blower-door method or the tracer gas method.

- Blower door tests measure the tightness of the structure and the integrity of the duct runs installed in an unconditioned space, but blower door tests do not directly measure the infiltration rate. (Efforts have been made to associate blower-door leakage measurements with infiltration rates, but some correlation procedures are not reliable. In this regard, the guidance published in the *ASHRAE Handbook of Fundamentals* has the most credibility.)

- Tracer gas tests measure the natural infiltration rate associated with a structure because the dilution rate of the trace gas depends on the infiltration rate. However, a certain level of skill is needed to perform this test because there is no way to control the wind velocity during the testing period.

- Any type of leakage test is difficult to perform because the result is affected by the wind velocity, the indoor and outdoor temperatures, the active number of exhaust and venting systems, the operating state (on/off) of the air handling blower, the operating state of the combustion equipment, and the position (open/closed) of the interior doors.

- If heating equipment draws combustion air from the surrounding space, a test should be performed to demonstrate that the combustion air requirements can be satisfied by infiltration. (Refer to Section 2 for information about this subject.)

4-8 Attic Ventilation

Attic ventilation is required by building codes and Federal Housing Authority (FHA) regulations because it moderates attic temperatures during the summer and reduces the amount of water vapor that is mixed with the attic air during the winter (if a vapor retarder is not used to keep moisture from migrating to the attic). Attic ventilation can be provided by various combinations of soffit vents, ridge vents, and gable-end vents. And, although they are not mandatory, mechanical devices (turbine ventilators and attic fans) can be used to increase the attic ventilation rate.

4-9 Exhaust Fans and Appliance Vents

Kitchen and bathroom exhaust systems, clothes dryer vents, and dishwasher vents are very desirable because they effectively control moisture and odors. These devices should always be installed, even if they are not required by code. (Some codes suggest that an exhaust fan is not required if a kitchen or bath features an openable window.)

4-10 Ducts In Unconditioned Spaces

If possible, a home should be designed so that all of the duct runs can be installed within the conditioned space. This is important because the leakage and conduction losses associated with runs in an unconditioned space degrade comfort and increase the design load on the HVAC equipment, the operating cost, the installation cost, and the demand load on the utility service.

If duct runs must be installed in an unconditioned space, they must be properly insulated and tightly sealed. Also note that structural cavities (stud spaces, joist spaces, and framed chases) should not be used as airways unless they can be completely sealed.

4-11 Codes and Regulations

Codes and regulations should be adjusted for the local climate. The Council of American Building Officials, Model Energy Code (CABO-MEC) and the American Society of Heating, Refrigeration, and Air Conditioning Engineers (ASHRAE) Standard 90.2 are two examples of climate-specific codes. (These documents represent the consensus opinion of a group of building industry agents. They do not have legal standing unless one or the other is sanctioned by a local or state law.) As far as the structure is concerned, these recommendations pertain to insulating requirements, sealing requirements, window and door performance, moisture control, R-values, U-values, ventilation requirements, combustion air requirements, exhaust fan requirements, and installation procedures.

4-12 Return on Investment

The costs of installing insulating materials, window assemblies, and doors escalate with the increase in performance. Therefore, once the minimum code requirements are satisfied and comfort issues are resolved, justification for an additional increment of insulation, a more efficient window, or a tighter fitting door has to be provided by an attractive reduction in operating costs. Of course, there are no general guidelines on this matter because every situation is unique.

- Installation costs depend on local conditions.
- Payback depends on local weather patterns.
- Payback depends on local utility rates.
- Payback depends on utility incentive programs.
- Each home has a different set of architectural features.
- Each home has a unique heating and cooling system.
- Lifestyle issues are hard to quantify.
- People use different types of payback criteria.
- Answers depend on the integrity and skill of the analyst.

Note that some codes and regulations are based on the results of a cost-benefit analysis. In this case, the design's efficiency can be optimized by simply complying with the minimum code requirements. (This assumes of course, that the calculation procedure used to produce the code is based on credible assumptions and a comprehensive set of alogrhythums.)

Section 5
Residential HVAC Systems

Residential comfort systems, whether intended for winter, summer, or year-round use, usually include primary energy conversion equipment, a distribution system, and controls. The primary equipment may be electrical or powered by some type of fossil fuel (natural gas, propane, or oil), the distribution system might be mechanical (air, water, or refrigerant-based) or electrical, and the controls are normally electrical or electronic. The energy conversion hardware may be designed to convert a fossil fuel or electrical power into heat (furnaces and electric resistance coils), or it may use electrical power or a fossil fuel to drive a machine that moves heat from a colder space to a hotter space (an air conditioner or heat pump). Systems also are classified by function (heating or cooling), by zoning capability (single-zone or multi-zone), and by other characteristics. A more comprehensive discussion of the various types of residential comfort conditioning systems is provided below.

5-1 System Classification

Residential comfort conditioning systems can be distinguished by function, distribution medium, zoning capability, type of fuel, heat sink, and/or heat source. Therefore, a complete description of a particular type of system will produce a string of identifying characteristics — for example, a single-zone, electric, earth-coupled, water-to-air heat pump, heating and cooling system. (Such comprehensive descriptions are seldom used. Normally, this system would be referred to as a single-zone, earth-coupled heat pump system, because this type of system typically features electric, water-to-air equipment that provides heating and cooling.)

Function

The functions normally provided by residential year-round comfort systems are heating, cooling, dehumidification, and air filtration, but they also can provide ventilation (outdoor air) and winter humidification. In some cases, two or more independent systems may be required to provide all of the desired functions.

Distribution Medium

The distribution medium can be a fluid or electrical wiring. Distribution fluids are used to transport heat between central energy conversion equipment and the conditioned spaces. Air, water, halogenated hydrocarbon refrigerants (F-22, for example), and steam can be used as distribution fluids. (Residential systems typically use one distribution fluid per system.)

- Air systems use a blower and duct runs to distribute hot or cold air to the conditioned spaces. The terminal equipment that serves the conditioned spaces may consist of locked-in-place hand dampers and fixed-geometry grilles and diffusers (constant volume system), modulating thermostat-controlled dampers and fixed-geometry grilles and diffusers (variable volume system), or locked-in-place hand dampers and modulating thermostat-controlled supply air outlets (variable volume system).

- Hydronic systems rely on pumps and pipes to distribute water (hot or cold) to the conditioned spaces. (If the home is very large, zoned cooling can be provided by a chilled water system.) The terminal equipment in the conditioned space may consist of baseboard convectors, a serpentine piping system installed under a floor, a fan-powered unit heater, or a fan-coil unit that serves a local duct system.

- Distributed direct expansion (DX) systems rely on refrigerant lines to supply refrigerant to remote terminal equipment located in or near the conditioned spaces. These lines connect a central condensing unit (which could be cooling-only or heat pump devices) and two or more fan coil units. Usually, the fan coil equipment discharges conditioned air directly into the occupied space, but some units are designed to accommodate a local duct system.

- Steam systems rely on boiler pressure and pipes to distribute steam to the heated spaces. The terminal equipment in the conditioned space may consist of a radiator, baseboard convector, unit heater, or fan-coil cabinet.

- Electric heating systems distribute power to local heating devices. These devices include baseboard convectors, heating coils embedded in the floor or ceiling, unit heaters, and fan-coils.

Zoning

HVAC systems can be classified as single-zone systems or multi-zone systems. A multi-zone system may accommodate as few as two zones or it could provide local control for every room in the house.

Fuel

The fuel options for heating equipment normally include natural gas, oil, propane, and electricity. Residential cooling equipment and heat-pump equipment are usually driven by an electric motor, but engine-driven equipment (that burns natural gas) also is available.

Heat Source and Heat Sink

Cooling equipment may be air-cooled or water-cooled. Air-cooled equipment expels heat directly to the outdoor air. Water-cooled equipment expels heat to the water that flows through the condenser.

Heat pump equipment is classified by the fluids that flow through the heat exchange devices. By convention, the fluid that flows through the compression-cycle equipment is named first (air-source or water-source) and the fluid that flows through the blower-coil is named second. The most common types of heat pump equipment are classified as air-to-air and water-to-air. Water-to-water and air-to-water heat pumps also are available.

Water source heat pump systems are identified by the characteristics of the water supply system. If well water or pond water makes one pass through the refrigerant-to-water heat exchanger, the system is classified as an open system. If the same water continuously circulates through a closed-loop piping system buried in the ground or submerged in a pond, the system is described as a closed-loop system or an earth-coupled system.

5-2 System Comparison

Figure 5-1 compares the characteristics of the most common types of residential heating and cooling systems. These comparisons are made on the basis of the distribution medium because it is the only mutually exclusive attribute.

Characteristics of Common Residential Comfort Systems				
	Distribution Medium			
	Forced Air	**Water**	**Electricity**	**Refrigerant**
Location of Equipment	Central	Central	Distributed Self-Contained Units	Central Compressor(s) Distributed Fan Coils
Primary Functions	Heating, Cooling, Dehumidification	Predominately Heating, Cooling is Possible	Heating, Cooling, Dehumidification	Heating, Cooling, Dehumidification
Air Filtration	Viscous Media Filters or Electronic Filters With Air Handling Unit	Need Separate System or Fan Coil Unit(s)	Filter in Local Units	Filter in Local Units
Provide Outdoor Air	Optional	Need Separate System or OA Fan Coil Unit(s)	Need OA Damper in Local Unit	Need OA Damper in Local Unit
Winter Humidification	Optional	Need Separate System	Need Separate System	Need Separate System
Cooling Equipment	Direct Expansion Coil	Water Chiller	Direct Expansion Coil	Direct Expansion Coil
Heat Pump Options	Air-to-Air Water-to-Air	Water-to-Water Air-to-Water	Air-to-Air	Air-to-Air
Heating Equipment	Furnace Electric or Fuel-fired	Boiler	Electric Resistance or Heat Pump	Electric Resistance or Heat Pump
Primary Fuel Options	Gas, Propane, Oil, and Electricity	Gas, Propane, Oil, and Electricity	Electricity	Electricity
Zoning Capability	Central 1-Zone and VAV Multi-zone	Many Zones or Single Zone	Many Zones	Many Zones
Terminal Devices	Diffusers, Grilles Registers, and Balancing Dampers	Baseboard Convector Unit Heater Fan Coil Unit Radiant Heating Pipes	Baseboard Heaters Radiant Cables Window Units PTAC & PTHP Units	Wall and Floor Mounted Blower Coil Units
Distribution System	Blower and Duct Runs	Pump and Pipe Runs	Electrical Wiring	Refrigeration Piping Plus Electrical Wiring
Operating Controls	Central Thermostat (on , auto, off) or Zone Thermostats For VAV Dampers With Central Panel (on , auto, off)	Zone Thermostats for Local Water Valves Plus Limit Control or Central Thermostat for Boiler or Chiller (on , auto, off)	Local Thermostat (on, off, or staged)	Zone Thermostats for Local Fan Coil Units Plus Microprocessor for Central Compressor (on , auto, off)

Figure 5-1

5-3 Central Forced Air Systems

A central forced air system is the most versatile of the four system types (see Figure 5-1) because it can be designed to provide any combination of the six comfort conditioning functions (heat, cooling, dehumidification, winter humidification, filtration, and ventilation) along with zone control (if desired). These systems also offer the greatest variety of primary fuel conversion devices (electric resistance, refrigerant-based, and fuel burning) and they are compatible with the most common fuels (natural gas, propane, oil, and electricity).

In regard to indoor air quality, forced air systems feature a circulating blower that can provide a continuous flow of air through every room. This is important because some air movement is necessary for the occupant's comfort (see Section 1). And if required, a supply of fresh air can be easily routed into the return-side of the duct system, as demonstrated by Figure 5-2. (This is an especially attractive feature when mechanical ventilation is required to meet an air quality standard.)

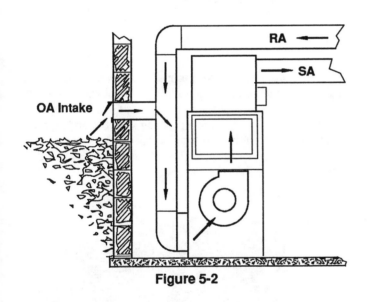

Figure 5-2

Since a forced air equipment package can be located at a point remote from frequently occupied rooms, the system designer has control over objectionable mechanical noise generated by a blower, compressor, condensing fan, or burner. Air-way attenuation also can be used to control equipment noise. (Duct liner or duct board is very desirable near the blower, and lined els provide a considerable amount of attenuation.) Of course there is also a possibility that noise could be generated within the duct system, but this should not be a problem if the duct runs and air outlets are sized in accordance with the procedures that are presented in **Manual D** and **Manual T**.

The disadvantages associated with forced air systems pertain to the duct runs and the terminal devices. Since they take up

a considerable amount of space, it may be difficult to route the duct runs through the structure without encroaching on the living space. And, if they are installed outside of the thermal envelope (in an attic or open crawl space, for example), they must be thoroughly sealed, insulated, protected by a vapor retarding material (for humid climates), and tested for leakage. In addition, the entire distribution system has to be balanced. And, if duct runs are installed in an accessible area, there is a possibility that they will be damaged by people, pets, or nesting animals.

There also are problems associated with the various types of supply air outlets. If the outlets are installed in the floor, there could be a conflict with one or more pieces of furniture or some of the window shading devices. If they are installed in a high sidewall or ceiling location, the temperature cannot be controlled at floor level when the room is built over a slab or exposed floor. (There are no "cold floor" problems associated with floors that are buffered by a heated basement.) Other problems attributed to supply outlets are caused by sizing and selection errors. Refer to **Manual T** for more information about supply outlet sizing procedures.

The furniture conflict problem also applies to low returns; otherwise it makes no difference whether the return is located near the floor or near the ceiling. Any of the other problems that are attributed to returns are caused by poor design. (Returns must be sized correctly, and a return or transfer grille is required for every room that has an interior door.)

Central Heating Equipment
The central heating equipment could be fuel-fired, or electrically powered, or it could be a dual-fuel unit. If a furnace is used, it could burn natural-gas, propane, or oil. In this case, the blower and filter are normally included with the furnace.

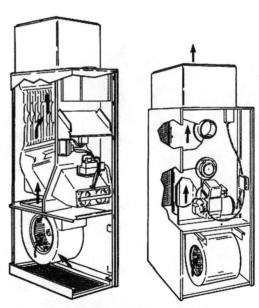

Gas and Oil Furnaces

If the central heating equipment is exclusively electric, it could be an electric furnace, an air-to-air heat pump, or a water-to-air heat pump. If an electric furnace is used, the blower, filter, and resistance heaters will be packaged in the same cabinet. If a split, air-to-air heat pump is used, a refrigerant coil, the blower, a filter, and a supplemental heating coil will be contained in the indoor package and the rest of the refrigeration-cycle machinery will be contained in the outdoor unit. If a single-package unit is installed, the air handling components will be packaged with the refrigeration-cycle machinery in an outdoor cabinet, and two trunk ducts (supply and return) will pierce the thermal envelope.

and filter will be part of the furnace package and the heat pump equipment will consist of an indoor refrigerant coil (mounted on the furnace) and the outdoor unit.

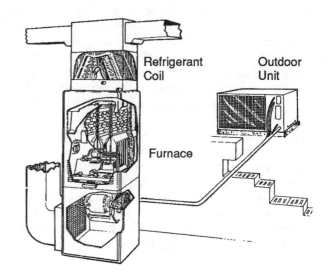

Furnace With Refrigerant Coil

Refrigerant to Air Heat Exchanger

Air-Handling and Refrigeration Cycle Machinery Packaged in One Indoor Cabinet

Water to Refrigerant Heat Exchanger

Water-to-Air Heat Pump

Hybrid equipment is unique because it features an open burner located below the refrigerant coil that is packaged with the outdoor unit. In this configuration, the burner heat is captured by the refrigerant and pumped (along with the compression cycle heat) to an indoor air handler that contains a refrigerant coil, the blower, and a filter. (See **Manual H** for more information about dual fuel equipment.)

Unit Installed Outdoors; Supply and Return Ducts Penetrate the Structural Envelope

Single-Package Unit

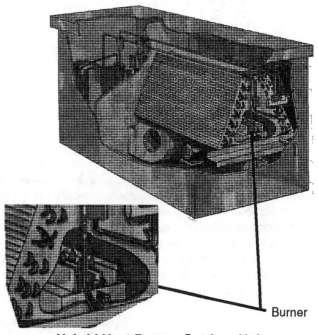

Burner

Dual-fuel equipment also is available. This equipment could consist of a split-system heat pump and a fossil fuel furnace, or it could be a patented hybrid system. If a split-system heat pump equipment is supplemented by a furnace, the blower

Hybrid Heat Pump - Outdoor Unit

Central Cooling Equipment

Central cooling systems could feature cooling-only equipment, an air-to-air heat pump, or a water-to-air heat pump. If cooling-only equipment is installed, it could consist of an outdoor unit and an indoor refrigerant coil is mounted on a furnace cabinet (electric cooling, fossil fuel heating), or it could be split system that features a cooling-only air handler containing a refrigerant coil, blower, and filter. Central, single-package cooling equipment designed to be installed outdoors also is available. (Outdoor packages require roof or wall penetrations for the supply trunk duct and the return trunk duct.)

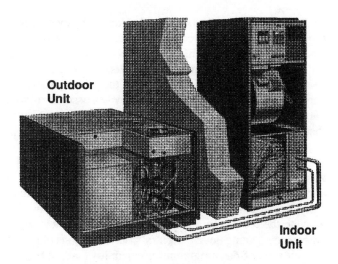

Outdoor Unit

Indoor Unit

Split Air-to-Air Equipment

Engine Driven Equipment

Equipment manufacturers, in cooperation with the natural gas industry, have developed an engine-driven heat pump. This equipment features an outdoor package that contains the compressor and the engine. The outdoor package also contains hardware designed to reclaim heat that would otherwise be discarded with the engine exhaust. The indoor unit contains the conventional air handling devices (refrigerant coil, blower, and filter), but supplemental heat is provided by a hot water coil instead of the traditional electric resistance coil. This water-to-air coil is supplied with water that is heated by the engine exhaust, and when necessary, a backup, gas-fired heating element.

Central Cooling Unit With Hot Water Coil

Some homes are being heated by a domestic gas-fired water heater. This type of year-round comfort system features an outdoor cooling-only condensing unit, an indoor fan-coil unit, a hot water coil (installed in series with the indoor cooling coil) and a domestic water heater.

The primary advantage of this system is related to the size of the gas furnace. If a home needs a significant amount of cooling and a moderate amount of heating, the furnace size

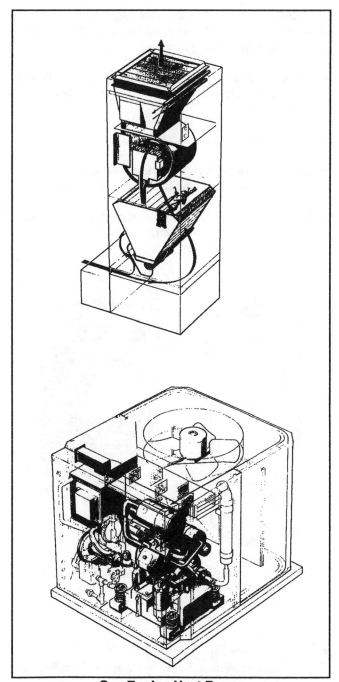

Gas Engine Heat Pump

will be dictated by the cooling load because the blower CFM must be compatible with the cooling coil air flow requirement. But, when the furnace size is based on the cooling season air flow requirement, it usually has too much heating capacity. This problem can be avoided with the domestic water heater system because the blower coil can be sized for the cooling load and the capacity of the hot water system can be matched to the heating load. (This type of heating-cooling sizing problem also can be solved by installing a heat pump, but the viability of this option depends on the cost of gas and the cost of electricity.) There also are secondary advantages associated with using one piece of equipment to perform the water heating and space heating functions.

• Domestic water heaters, which are usually on a stand-by status, are more efficient when they are subjected to a steady load.

• Water heater vents are relatively easy to install.

• During the heating cycle, the supply air feels warm, but it is not hot enough to cause stratification problems.

• The heating coil is never gets hot enough to burn air borne particles (fewer odor problems).

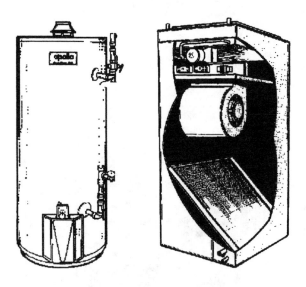

Domestic Water Heater and Fan Coil Equipment

Optional Equipment
The optional equipment found in central air handling units usually involves an air filter or some type of humidifier. A more efficient filter (based on dust spot efficiency) is desirable because standard filters cannot capture the small particles normally entrained in the indoor air. (If an accessory filter is installed, the pressure drop across the filter must be compatible with the performance of the blower.) A humidifier is desirable during very cold weather because dry air causes static electricity problems, physical discomfort (more evaporation from the skin surface), and respiratory discomfort (dry sinuses).

Supervisory Controls
Single-zone, forced air systems are controlled by a central thermostat that monitors the operation of the fuel conversion equipment and the blower. This thermostat may be designed to control a single-function system (heating-only or cooling-only), or it may be designed for year-round operation (heating and cooling). In any case, the capabilities of the thermostat must be compatible with the capabilities of the equipment. For example, an automatic change-over capability is appropriate for year-round thermostats and a sequencing capability is required if the equipment capacity can be activated in two or more stages. (The second stage of a two-stage thermostat is commonly used to cycle the electric resistance coil that supplements the heating capacity of a heat pump unit when the outdoor temperature falls below the balance point. Two-stage thermostats also are required when the system features multi-speed equipment.)

Multi-zone systems may be controlled locally by independent room thermostats, or by local thermostats that are supervised by a central logic panel. Electric baseboard heating systems, for example, can be controlled by a collection of autonomous heating-only thermostats. By contrast, the primary fuel conversion equipment and the air-side devices associated with variable volume systems are controlled by a collection of local thermostats and a central microprossor.

Also note that the central thermostats supplied with single-zone systems and the central control panels that operate variable volume systems usually include one or more manually operated switches. These switches are normally used to activate the blower on a continuous basis or to energize an emergency heating device. Central control stations may also provide time-of-day programming capability that can be used to implement day-of-week set-up and set-back schedules. (Programmable devices normally include a manual over-ride switch that can be used at the discretion of the occupant.)

A humidistat will be required if a heating-cycle humidification device is added to the air distribution system. This control, which is independent of the temperature controls, can be located in the conditioned space or near the air handler. (If the humidity control is located near the air handler, a sensor in the return air duct can be used to monitor the moisture content of the return air.)

Distribution System
The diagrams on the following page show that residential supply-air distribution systems can be configured in a simple extended plenum arrangement (reducing or non-reducing), a primary-secondary trunk arrangement (rigid or flexible), or in a radial pattern. The complimentary return system could be as sophisticated as the supply system (return installed in every room) or it could be more basic (returns in selected areas supplemented by transfer grilles). Duct runs could be installed above the ceiling, in a structural cavity, in a crawl space, or in a basement, or they could be buried below a slab floor. Rigid duct runs normally are fabricated from galvanized metal or rigid fibrous board material. Flexible, spiral wire ducts also are popular. Comprehensive information about the different types of air distribution systems and air-side devices can be found in **Manual D**, Section-1 and -2.

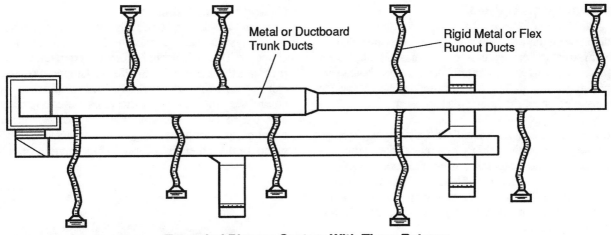

Metal or Ductboard Trunk Ducts

Rigid Metal or Flex Runout Ducts

Extended Plenum System With Three Returns

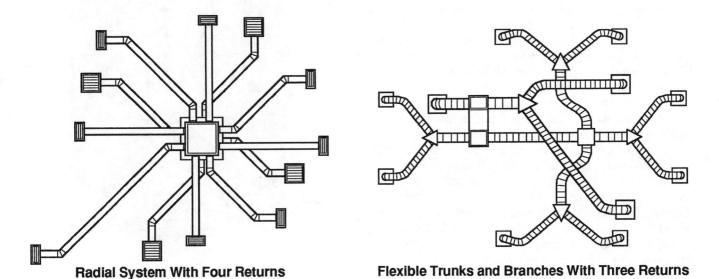

Radial System With Four Returns

Flexible Trunks and Branches With Three Returns

Primary-Secondary Trunk System With Four Returns

5-4 Circulating Water Systems

The majority of the circulating water systems installed in residential structures are heating-only systems. Usually, these systems feature convective terminal devices (baseboard fin-coil units, for example), but radiant panel designs (hot water tubing installed under the floors) also are used.

The primary advantages associated with free convection heating systems pertain to their ability to provide zoned temperature control and their tendency to warm the interior surfaces of the outside walls and floors (which reduces the radiant heat loss from exposed skin surfaces). And, there is an additional benefit associated with the minimal space required for the distribution piping.

Unfortunately, a free convection heating system cannot satisfy all the important comfort conditioning requirements. For example, this type of system will not filter indoor air or provide outdoor air for ventilation. One solution to this problem involves installing an independent, ducted, forced-air ventilation system equipped with a heat recovery device (at substantial expense, of course).

Alternatively, the ventilation and filtration functions can be furnished by a design that features a ducted forced-air heating system. For example, outdoor air and recirculated air could be routed through a blower cabinet equipped with a hot water coil and a filter. Forced air systems also are useful for providing winter humidification. And, if year-round comfort is desired, this system could be used to provide the cooling and dehumidification functions. In this case, a chilled water coil would be installed in the fan cabinet and a water chiller would be added to the roster of central equipment.

Applications

Convective heating systems are compatible with cold-climate homes that have enough infiltration to satisfy the minimum fresh air requirement (for humidity control and air quality), but they may not be the best choice for a tightly sealed home or for homes that require year-round conditioning. (As noted earlier, homes that do not have enough infiltration could benefit from some type of independent ventilation system.)

Air-water systems are compatible with any type of home because they can provide acceptable air quality, year-round temperature control, summer dehumidification, and winter humidification. However, for most homes, this type of system is less desirable than the more common types of central all-air systems (featuring a furnace, air conditioner, or heat pump). The exception is a very large home. In this case, large-scale zone control can be provided by two or more air handling systems, a central boiler, a central water chiller, and distribution piping. And, if desired, one or more air handling systems could be equipped with air-side hardware that allows variable-volume zoning on a room-by-room basis. (These types of air-water systems are commonly used in commercial buildings.)

Central Heating Equipment

Hot water boilers can be fired by natural gas, propane, or oil; or the system water can be heated by electric resistance elements. This equipment is normally trimmed to operate at pressures that are less than 30 PSI and temperatures that range from about 140°F to about 240°F (depending on the limit control setpoints), but in practice, a narrower operating range of 170°F to 190°F is typical. It also is possible, but not very common, to substitute an air-to-water or a water-to-water heat pump for the boiler (see **Manual H** for more information).

Small Gas-Fired Boiler

The water temperature rise (TR) across a boiler depends on the output capacity (BTUH) and the GPM flowing through the boiler. This temperature rise can be computed by using the water-side equation.

$$TR = \frac{Output\ BTUH}{500 \times GPM}$$

The temperature rise across a boiler is limited by the potential for causing thermal shock. Cast iron boilers can tolerate a 25°F to 30°F temperature rise. For steel boilers, the temperature rise should be limited to 20°F. Note that the temperature rise across the boiler and the temperature drop across the piping circuit must be compatible with each other.

Central Cooling Equipment

Chilled water systems are not normally used to cool residential structures, but if the home is very large, an electric water chiller could be used to provide summer conditioning. Air-cooled packages as small as 5 tons (rated at 95°F outdoor air temperature) are available from some manufacturers.

Terminal Equipment

Heat can be delivered to a conditioned space by a strip of baseboard radiation, a convection coil mounted in a cabinet (recessed, wall mounted, or floor model), a fan coil unit, or serpentine piping circuit installed below the floor (radiant heat). Cooling can be provided by a small air handler and a duct system or by unitary fan-coil equipment. (Air handler and fan-coil equipment also provides an opportunity to filter the air.)

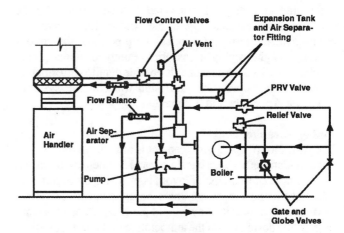

Figure 5-3

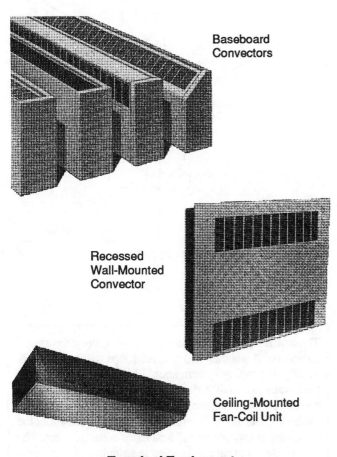

Baseboard Convectors

Recessed Wall-Mounted Convector

Ceiling-Mounted Fan-Coil Unit

Terminal Equipment

Hydronic Accessories

Water-side fittings and accessories are required to protect the equipment and control the system. Piping hardware includes check valves, strainers, drains, shut-off valves, control valves, relief valves, balancing valves, gauges, and flexible connections. Hydronic accessories include the expansion tank, pressure reducing valves, air-separators, and air-vents. Figures 5-3 and 5-4 shows how these devices might be installed in the piping plan.

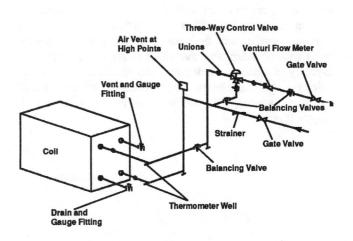

Figure 5-4

Expansion Tank

Water-side pressures must be maintained within certain limits during any operating condition, even when the system is idle. This is accomplished by installing an expansion tank. This tank controls system pressure by accommodating the volumetric expansions and contractions associated with changing water temperatures. The size of the expansion tank depends on the volume (in gallons) of the hydronic system, the operating temperatures, and the operating pressures. Hydronic specialties manufacturers provide comprehensive sizing and installation instructions in their engineering literature. Piping design manuals provide information about the volumetric properties of pipes, and HVAC equipment manufacturers provide information about the holding capacity of the water-side components.

Pumps

A pump must be selected to deliver the design flow (GPM) when working against the maximum resistance or head (which is measured in feet of water) that could be created by the piping circuit(s). This resistance is determined by the pressure drops associated with the primary fuel conversion equipment, the terminal devices, the pipe runs, the control valves, and the piping accessories. Pump manuals, piping manuals, and hydronics manuals provide detailed information about the procedures used to design water piping systems. (Also see Appendix 5 in this manual.)

Piping Arrangements

The series loop piping arrangement is often used for residential applications because it is simple and inexpensive to install. An example of this type of layout is provided by Figure 5-5, which shows how the water flows through a queue of terminal devices as it makes its way around the circuit. The disadvantage of this design is that the temperature of the water progressively decreases as it flows from device to device. This means that to provide the same amount of heat, a downstream device must have more fin area than an upstream device. However, this is not a problem if the total temperature drop across the loop can be limited to 20°F or less. (In this case the size of the terminal devices can be based on the average loop-water temperature.) Also note that there is no way to adjust the capacity of the terminal devices individually from the water-side (dampers could be used to provide some air-side control), and that the water-side pressure drop is proportional to the length of the loop.

loops are common. This type of piping arrangement is cost-competitive with a single loop design and it holds the temperature and pressure drops to acceptable limits. It also provides obvious opportunities for zone control. However, there still is no way to implement water-side control on a device-by-device basis.

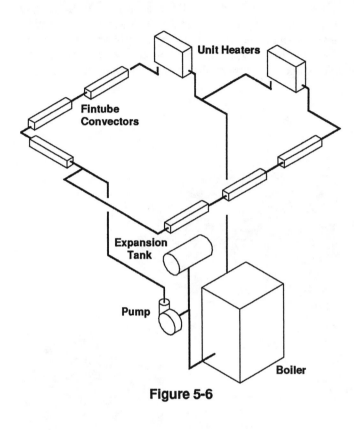

Figure 5-6

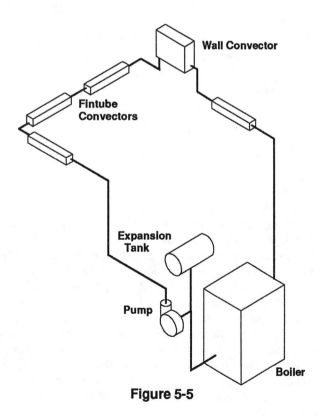

Figure 5-5

The total temperature drop around the loop provides a good indication of whether two or more loops will be required. Ideally, the piping system should be designed for a loop temperature drop of 20°F or less. This 20°F value will be compatible with the recommended temperature rise across the boiler, and all the terminal devices will have about the same heat output per unit of heat transfer surface area. The water-side equation can be used to calculate the temperature drop (TD) across a loop.

$$TR = \frac{Capacity\ of\ Loop\ Devices\ BTUH}{500\ x\ Loop\ GPM}$$

If the home is large, parallel loops provide a way to avoid the problems associated with installing a long, single-loop system. Figure 5-6 shows a two-loop design, but three or more

A one-pipe system allows a terminal device to be controlled by a two-position valve. Figure 5-7 (on the next page) shows that this arrangement uses special "one-pipe tees" to establish a branch loop for each terminal device. (This hardware is proprietary. Manufacturer's performance data must be used to select the product that produces the desired branch flow.) Because branch risers, drops, tees, and valves are required, this system will cost more to install than a series loop system, and it is subject to the same temperature and pressure drop limitations. Also note that Figure 5-7 shows only a single loop system. If the home is large, parallel loops can be used to limit the loop temperature and pressure drops, and to provide zone control.

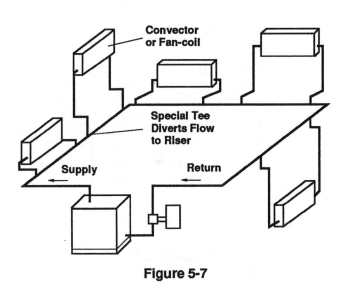

Figure 5-7

If a two-pipe system is installed, it could feature a direct or reverse return piping layout. Figure 5-8 shows a direct-return arrangement. Notice that the distance the water must travel from the central equipment to a terminal device and back to the central equipment is different for every device. This means that the pressure drop associated with the piping runs is different for every device. The smallest piping loss is generated by the circuit that serves the closest device, and the largest piping loss is produced by the circuit that serves the furthest device. Therefore, this piping arrangement is inherently unbalanced.

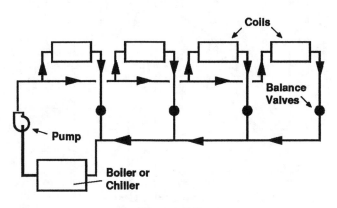

Figure 5-8

Figure 5-9 shows a reverse-return arrangement. In this case, the distance that the water must travel from the central equipment to a terminal device and back to the central equipment is the same for every device. This means that the pressure drop associated with the piping runs is the same for every device. Therefore, this arrangement produces balanced piping system pressure losses.

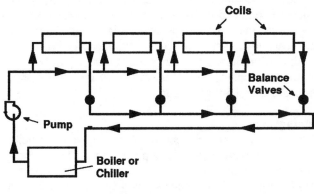

Figure 5-9

The advantage of direct-return arrangement is that less pipe is required for the return run. The advantage of reverse-return arrangement is that the piping losses are balanced. The disadvantage of direct-return arrangement is that balancing valves must be installed in the branch risers. The disadvantage of reverse-return arrangement is that more pipe is required for the return run. The reasons for selecting a direct- or reverse-return arrangement depend on the relationship between the piping losses and the pressure drops across the heat transfer devices.

• If the pressure drops across the remote devices are equal, balancing valves are not required, providing that the piping losses are balanced by a reverse-return layout.

• The pressure drops across the remote devices are not important if they are considerably smaller than the piping losses. (Fin-tube radiation and cabinet convectors may have relatively small water-side pressure drops.) In this case, balancing valves are not required, providing that the piping losses are balanced by a reverse-return layout.

• If the pressure drops across the terminal devices are not equal, the system will be inherently unbalanced, so there is no advantage to the reverse-return layout. A direct-return arrangement with balancing valves is appropriate in this situation.

• A direct-return arrangement with balancing valves is appropriate when the pressure drops across the terminal devices are large in comparison to the piping losses. (Fan-coils and unit heaters have relatively large pressure drops.)

The advantage of using a two-pipe system is that the same water temperature (ignoring pipe losses) can be maintained at the entrance to each piece of terminal equipment. Therefore, the full heating or cooling potential of the supply water is available at the entrance to every heat transfer device. And. since each terminal device can be independently controlled, these systems are much more versatile than the series loop and one-pipe systems. (They can serve any number of zones, they can be used for winter heating or summer cooling, and fan-coil units can be mixed with convection devices.) The disadvantages of the two-pipe system are associated with the installation cost and the ability to provide simultaneous heating and cooling. (Simultaneous heating and cooling is not normally required for homes. See **Manual CS** for more information about 2-, 3- and 4-pipe systems.)

Zoning

Figure 5-10 shows how control valves or circulating pumps can be used to zone parallel loop systems. As far as performance is concerned, pumps are preferable to valves because they ensure that each piping loop will operate as an independent circuit. This means that the pumping load (which is characterized by the flow rate and operating head) of one circuit will not be affected by the load fluctuations in the other circuits. But, as far as installation costs are concerned, multiple circulating pumps are more expensive than a set of control valves.

If a single pump and a set of two-way control valves are used for zone control, an unstable pumping load could be created as the zone valves open and close. But, if three-way mixing or diverting valves are used, the flow rate at the pump will be fairly constant because unneeded supply water will be bypassed to the return main. Figure 5-11 shows how the different types of control valves are installed in the piping circuit.

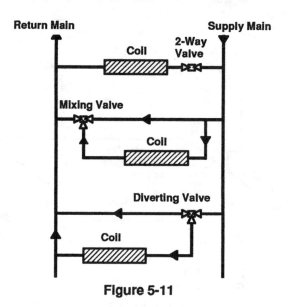

Figure 5-11

Controlling the Capacity of a Terminal Device

The simplest and least expensive way to control the capacity of a convection device or series of convection devices installed in a loop is to use two-position control (valve open or closed, pump on or off). But, some type of modulating control is more appropriate for fan coil equipment, and it maximizes the comfort that can be provided by convection equipment.

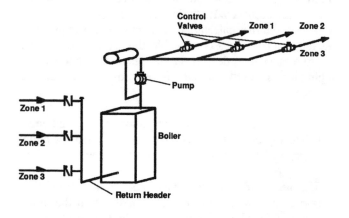

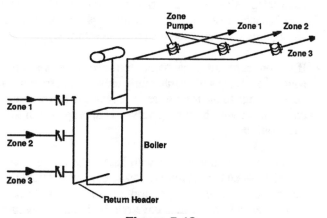

Figure 5-10

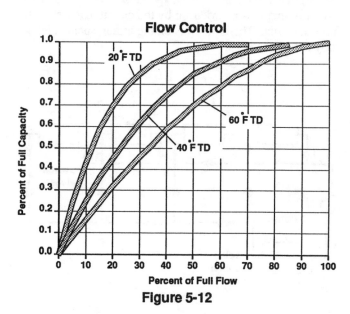

Figure 5-12

The capacity of a terminal device (fan coil or convective fixture) can be modulated by adjusting the flow through the coil (see Figure 5-12) or by changing the temperature of the water entering the coil (Figure 5-13). In this regard, temperature adjustment has an advantage over flow adjustment because a linear response is associated with temperature adjustment and a non-linear response is associated with flow adjustment. However, flow control valves are quite common because:

- A single water temperature may not satisfy the loads associated with each zone or piece of terminal equipment.

- Valves are simple and inexpensive to install.

- Equal percentage valves can be used to compensate for the nonlinearity associated with flow control.

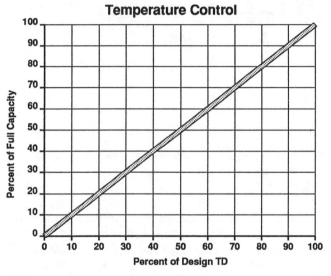

Figure 5-13

Supervisory Controls
The boiler, the pump or pumps, and the valves may be controlled by a single central thermostat or a set of zone thermostats. In either case, a system of relays and interlocks will be required. Refer to hydronic system design manuals for more information on this subject.

Boiler Controls
Boiler capacity control is important. Two-stage burners or modulating burner controls are preferred for applications that have large heating load fluctuations. Study the manufacturer's literature, as it pertains to controls and accessories. Specify the necessary burner operating controls, safety controls, and trim.

- Burner capacity controls — on-off, two-stage, modulating
- Ignition and flame supervision controls
- Master-submaster temperature reset controls

- High- and low-limit controls
- Pressure temperature gauge
- Pressure relief valve
- Drain cock.

5-5 Radiant Heating Panels

Since radiant heating panels transfer heat to the occupants directly, they have the ability to provide superior comfort during the heating season. In fact, the level of comfort associated with a forced air system and a 75 °F room air temperature can be duplicated by a radiant-floor system and a 65 °F room air temperature. (The radiant heating effect has a stronger influence on the physiological response of the occupant than the convection heating effect.)

> Radiant floor heating provides optimum heating-season comfort in high-bay rooms, great rooms and other areas that have high ceilings. If the rest of the home is served by a forced-air system, it may be possible to use a domestic water heater to generate the required supply of hot water. (The concept of using a domestic water heater for space heating service was introduced earlier in this section.)

Indoor Air Quality
Radiant heating systems are compatible with cold-climate homes that have enough infiltration to satisfy the minimum fresh air requirement (for humidity control and air quality). If a home is too tight, air quality can be maintained by installing some type of independent ventilation system. Also note that when year-round comfort is required, the cost of installing an air system for cooling and a hot water system for heating will be larger than the cost of installing an all-air heating and cooling system.

Radiant Panel Construction
A radiant panel can be provided by installing a hot water piping circuit or an electric resistance cable circuit in the floor assembly, or by installing an electric resistance cable circuit above the surface of the ceiling. In this regard, the choice between butelene water pipes and electric heating cables will be influenced by the complexity of the installation work, the cost of installing the system, and the cost of the fuel used to power the system.

If the floor assembly is used to create a radiant panel, the heating elements can be embedded in a concrete slab or in a gypsum membrane poured over a wood sub-floor, or the elements can be laid between the top of a gypsum membrane and the bottom of a sub-floor. If ceiling heat is desired, electric cable can be embedded in plaster under a sheet of lath. Regardless of the location of the panel, a generous amount of

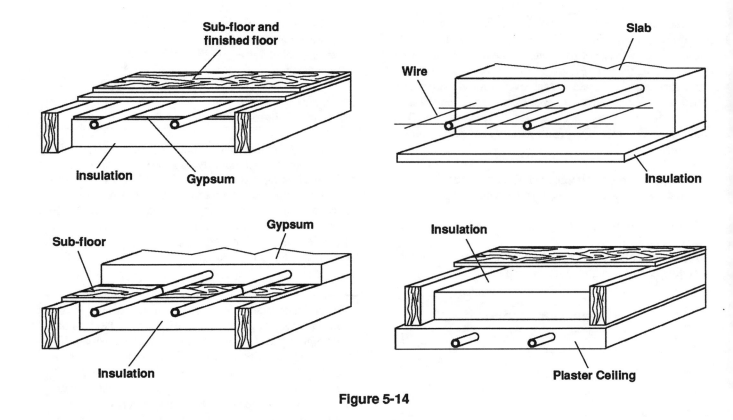

Figure 5-14

insulation must cover the cold-side of the assembly. Figure 5-14 illustrates the details of the various panel-construction options.

When a floor panel is covered by wood flooring, the heat tends to dehydrate the wood, and this causes shrinkage in the flooring material. This shrinkage is undesirable because it creates gaps between the strips of flooring material. This problem can be minimized by making sure that the moisture content of the flooring material is 8 percent or less at the time of installation. It also helps to use narrow flooring strips, and the problem can be minimized if laminated blocks of flooring are arranged in an orthogonal pattern. It also is important to make sure that the design does not allow the temperature of the flooring material to exceed 90°F.

If tile finishes the floor assembly, the seams in the tile should be coincident with the expansion seams in the substrate. In addition, an elastic material should be used to fill between these particular tiles.

Carpet reduces the ability of the panel to radiate heat to the surrounding surfaces. If carpet is laid on top of the floor assembly, it should have a minimal thermal resistance, and it should be glued to the floor.

Circuiting
Since electrical heating cables deliver the same amount of heat per foot of wire, repetitive circuiting patterns can be expected to generate a uniform flux across the panel. Or, if more heat is desired near the outside walls, the geometry of the pattern can be biased to deliver a greater flux at the edge of the pattern.

Circuiting a length of water pipe is more difficult because the temperature of the water decreases in the direction of the flow. This means that if an unfolded length of pipe winds its way across the room, or spirals around the room, more heat will be delivered to one area of the room. This behavior can be corrected by routing a folding pipe through the circuiting plan. This way, a uniform flux will be delivered to the room because the segments carrying the warmest water will be located next to the sections transporting the coolest water. It also is possible to design folded and unfolded circuits that deliver a larger flux along the outside walls. Figure 5-15 (see next page) provides examples of common piping circuits.

Operating Parameters
The length of electric cable that must be installed depends on the heating load and the heating capacity of one foot of cable. Refer to manufacturer's data for capacity information and installation instructions.

The length of water pipe required to neutralize the heating load depends on the heating load, the entering water temperature, and the temperature drop across the circuit. In this regard, the entering water temperature should be compatible with the 90°F limit on the floor temperature, and the temperature drop across the circuit should not be more than 20°F. Also note that the circuit length also is limited by the water-

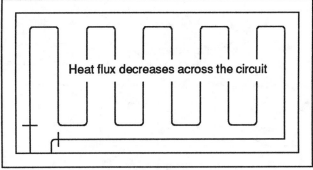

Unfolded Serpentine

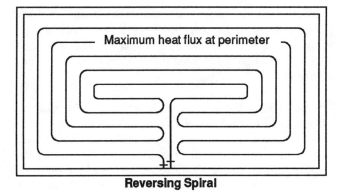

Reversing Spiral

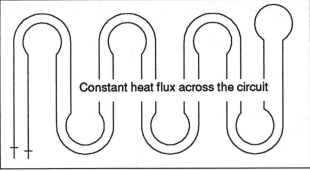

Folded Serpentine

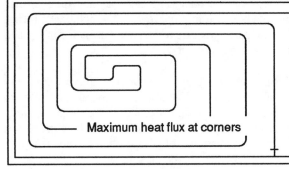

Folded Spiral

Figure 5-15

side pressure drop associated with the route, which in turn depends on the diameter of the tubing. The following guidelines can be used for rough estimates.

- Use 1-1/2 feet of 1/2" tubing per square foot of floor
- 1/2" circuits should be less than 275 feet long
- Use 1 foot of 3/4" tubing per square foot of floor
- 3/4" circuits should be less than 475 feet long.

Also note that after a water system is installed, it must be pressure tested, flushed and purged (no air in tubing, headers, or manifolds). The circuits also must be balanced to obtain the desired pressure drop and temperature drop.

Control
Electric systems can be controlled by a thermostat that interrupts the power supply when the thermostat is satisfied. Zone control is easily achieved by installing one power-circuit and one thermostat per zone.

Hot water systems require central controls and local controls. The central control at the hot water generator (boiler or water heater) must maintain the supply water temperature set-point. In this regard, performance can be improved by using a master-submaster control that schedules the supply water set-point in accordance with the outdoor air temperature (set-

point rises as outdoor temperature drops). Local control for a room or zone can be provided by a thermostat that controls the flow through a circuit valve (two- position or modulating).

5-6 Distributed Systems

Small refrigeration-cycle packages and resistance heating devices can used to supplement the capabilities of a central system (forced air or hot water), or this class of equipment could be used to condition an entire home. Simple examples of supplemental applications are provided when a window cooling unit adds a function that cannot be delivered by a central hot water heating system, or when an electric baseboard fixture is used to partially heat a room that needs to be zoned. As far as whole-house applications are concerned, the classical example of a distributed system is provided by a dwelling that is heated by independently controlled, electric baseboard fixtures.

Window air conditions and electric baseboard fixtures are not the only types of distributed devices at the designer's disposal. Packaged terminal air conditioners (PTAC units) could be used to cool and dehumidify a large open area, or a packaged terminal heat pump (PTHP unit) could be used to satisfy the cooling and heating needs of a similarly sized space. Alternatively, a ductless, split-coil air conditioner or split-coil heat pump could be used to condition one or more rooms, depending on the number of indoor coils attached to an outdoor unit. (Refer to Section 6 for more information about distributed equipment.)

One disadvantage to this approach to space conditioning is a limited ability to control the quality of the indoor air. In this regard, refrigeration-cycle equipment has more capability than resistance heating devices. But even if refrigeration-cycle equipment is installed, the efficiency of the filter will be relatively low and there may be no provision for mechanical ventilation. Or, if the equipment is designed to provide a supply of outdoor air, the flow is typically controlled by a manual, open-closed damper. Also note that maintenance requirements increase in proportion to the number of refrigeration-cycle packages, and remember that each cooling coil requires a condensate drain.

On the plus side, distributed systems are relatively simple to design and install (no duct system, integral controls), so selective use of small equipment packages or electric heating devices provide a simple solution to some zoning problems. There also is a potential cost-of-operation benefit if equipment packages are set up, set back, or shut down on a zone-by-zone basis.

5-7 Evaporative Cooling Systems

Under certain circumstances an evaporative cooling system can be used in place of a mechanical cooling system, or it could be used to reduce the peak load on the refrigeration-cycle equipment during the hottest days of the summer. The decision regarding the use of an evaporative cooling system depends on technical issues and economics. In this regard, the system must be able to maintain an appropriate level of comfort during any possible operating condition. If this is possible, the energy cost will be noticeably less than the cost of operating compression-cycle equipment.

Obviously, successful application of evaporative equipment requires an arid climate because cooling capacity is generated by evaporating water into a flow of outdoor air. This process can be illustrated on the psychrometric chart — as indicated by line A-B on Figure 5-16. Notice that as the air passes through the cooler it loses sensible heat and exits at a lower dry-bulb temperature, but it gains an equivalent amount of latent heat, producing a considerable increase in absolute humidity (the moisture content of a pound of air).

After the air has been cooled and humidified it is piped through an air distribution system and introduced to the conditioned space. Since this air is cooler than the indoor air, it can absorb sensible heat, and if it is dryer than the indoor air, it can provide some latent cooling. It follows that the temperature and humidity of the indoor air will depend on the supply CFM, the condition of the air leaving the evaporative equip-

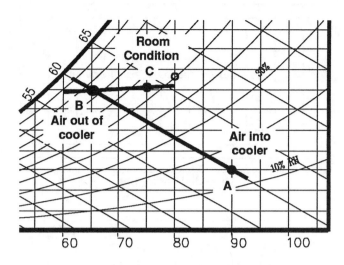

Figure 5-16

ment, the size of the sensible load and, the size of the latent load (which might be negative in a dry climate). This process also can be diagramed on the psychrometric chart, as demonstrated by Figure 5-16, line B-C. This diagram indicates that comfort conditions in the home will be maintained if point-C falls on the desired dry-bulb line, and if the relative humidity falls within the boundaries of the ASHRAE comfort chart (see Section 1).

Also note that an evaporative system is a one-pass system. (If processed air is recirculated through the evaporative equipment, it will become saturated after four or five passes. This in turn, will cause a loss of sensible cooling capacity, and, because the air is saturated, the indoor humidity would be excessive.) Refer to Section 6 for more information about evaporative equipment.

5-8 Reclaim Ventilation Systems

Some tight homes have been equipped with an independent ventilation system. As illustrated by Figure 5-17 (next page), these systems consist of a heat recovery ventilator (HRV), an outdoor air duct, a supply duct system, a return duct system, and an exhaust duct. The key components in the HRV include an air-to-air heat exchanger and two fans. One fan moves outdoor air through the heat exchanger and delivers the processed air to the conditioned space via the supply duct system. The second fan draws an equivalent amount of air from inside the home (balanced system), routes it through the heat exchanger, and discharges it to the outdoors through the exhaust duct. Other components contained in the HRV could include a filter, drains, balancing dampers, an electric heater, a defrost control, and the operating controls. The filter is required to improve the quality of the ventilation air. The drains are required because condensation may occur at the intake-side of the HRV during the cooling season, or on the exhaust-side during the heating season. And, the electric heater could be used either to defrost the HRV if condensation occurs during cold weather, or to add heat to the ventilation air during extremely cold weather.

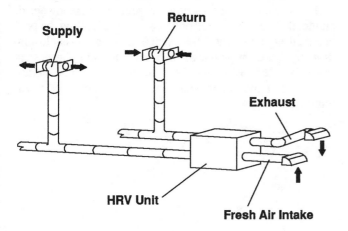

Figure 5-17

The disadvantages of this type of ventilation system pertain to the space required for the duct runs, supply outlets, and returns; the problems associated with mixing partially conditioned supply air with indoor air, and the cost of installing a complete air distribution system to deliver a relatively small amount of air to the conditioned space. However, these problems can be resolved if both sides of the HRV unit are piped to the return-side of a central forced air system, as indicated by Figure 5-18. This way the secondary distribution system is eliminated, all the air is filtered and fully conditioned before it reaches the conditioned space, and the installed cost is reduced. (The concept of installing a completely independent air distribution system (Figure 5-17) is compatible with tight, cold climate homes that do not have a central forced-air conditioning system.)

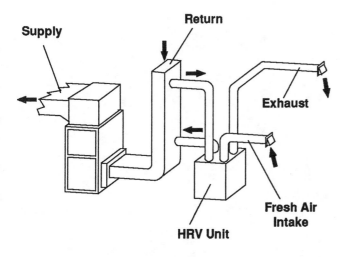

Figure 5-18

5-9 Hot Tubs and Swimming Pools

Homes that feature an indoor hot tub or swimming pool create an especially challenging problem for the builder and the HVAC system designer because moisture readily migrates through structural cracks and permeable surfaces, and it rapidly travels through duct systems. This means that rooms enclosing a hot tub or pool must be completely isolated, structurally and mechanically, from the rest of the home. If this water vapor is not contained and controlled, condensation can damage the structure (freezing condensation is especially destructive), fog windows, degrade comfort, precipitate biological growth (mold and mildew), or cause of musty smells and other types of air quality problems.

A second attribute of swimming pool and hot tub applications pertains to the amount of energy required to maintain acceptable temperature and humidity within the dedicated space. If the demand for energy is substantial, some type of energy recovery device, or system, could be justified by the potential reduction in operating cost.

As far as installation cost is concerned, the least expensive method that controls indoor humidity consists of a ventilation system that replaces humid indoor air with dry outdoor air. However, this approach is effective only when the absolute humidity (grains of water per pound of air) of the outdoor air is significantly less than the absolute humidity associated with the indoor design condition. This differential universally occurs during cold weather, but unless the home is located in an arid climate, the difference between the indoor and outdoor humidity diminishes during intermediate weather and disappears or reverses during hot weather. In other words, a ventilation system can be used if warm-weather humidity control is not required — which would make sense if there is an adequate amount of screened window and skylight area associated with the boundaries of the enclosure.

If a ventilation system is installed, the operating efficiency can be improved by including an air-to-air recovery device. This will reduce the amount of heat required to warm the flow of raw outdoor air. In this regard, it is important to understand that the recoverable energy consists of sensible heat and latent heat, and that the latent component represents a significant percentage of the total energy content. Therefore the cost-benefit offered by devices designed to recover sensible heat and latent heat should be compared to the cost-benefit provided by sensible-only units.

Since a ventilation-only system cannot control the indoor humidity during cool wet weather, mild humid weather, or hot humid weather; a mechanical cooling system, or a mechanical dehumidification system, is required for continuous humidity control. Therefore, when year-round conditioning is required, the designer should consider dehumidifier-water-heater heat pump equipment. This hardware is designed specifically to reclaim sensible and latent heat while controlling temperature and humidity during the moist adverse weather conditions. In addition, it uses some of the reclaimed heat to maintain the temperature of the pool water.

Also note that the standard load calculation and equipment sizing procedures documented in **Manual J** and **Manual S** do not apply to pool and hot tub applications. A comprehensive discussion of this subject can be found in pool and hot tub literature published by ACCA. Manufacturers of dehumidifier-water-heater heat pump equipment are another source of information. In this regard, the system designer must have a comprehensive understanding of all the issues and procedures associated with producing a viable design. If an individual is not comfortable with these concepts, the client should be advised to consult with designers or organizations experienced with this particular application.

In any case, designers and contractors should be prepared to deal with the expectations of an uninformed homeowner or builder that has considered only the cost of installing the pool or hot tub equipment and the cost of a conventional enclosure. In many cases the unbudgeted cost of providing structural isolation, a dedicated HVAC system, and heat reclaim equipment, and/or the anticipated increase in energy costs will scuttle the project.

Section 6
Primary Equipment

Forced air equipment, by a direct process, adds sensible heat to, or extracts sensible heat from, the indoor air. In most cases this energy transfer is accomplished by circulating the indoor air through an electric coil, a furnace heat exchanger, or a direct expansion refrigerant coil. In addition, direct-transfer equipment can be used to dehumidify the indoor air (circulation through a refrigerant coil), or to increase the moisture content of the indoor air (circulation through a humidifier).

Hot water heating systems (hydronic systems) and steam heating systems also rely on direct-transfer processes. But in this case, the energy transfer occurs in two steps. First, water is heated in the primary equipment (boiler), then this heat is transferred to the indoor air by a remote device, which could be a hot water coil located in an air handler, a baseboard (fin-tube) convector, or a wall-mounted convector.

Air handlers, baseboard convectors, and wall mounted convectors also are used with electric resistance heating systems. In this case, the heat that is generated by the resistance element is directly transferred to the indoor air.

Indirect heat transfer processes are associated with radiant panel heating systems. With these designs, electric resistance heating wires or hot water pipes are installed in or under a masonry floor, or in a plastered ceiling. This way, heat is transferred to the structural material, which releases some of this heat, by convection, to the indoor air, and radiates the balance to the surrounding surfaces (that eventually transfer the absorbed heat to the indoor air).

More detail about heating and cooling equipment and associated devices is provided here. This presentation is complimented by Section 5 of this manual (systems review) and by **Manual H** (heat pumps). Also refer to **Manual S**, for a comprehensive discussion of the sizing procedures that pertain to residential equipment.

6-1 Primary Equipment

The primary equipment used for residential applications could be a device that converts fuel into heat or a power consuming device that moves heat. Fossil fuel furnaces (natural gas, oil, or propane) and electric resistance coils are examples of heat generating devices; and the common types of refrigeration-cycle equipment (cooling-only packages, air-source heat pumps, and water-source heat pumps) are examples of heat moving equipment.

For the purpose of classification, residential furnaces are characterized by heating capacities that are less than 225,000

BTUH, residential boilers are limited to 300,000 BTUH, and residential refrigeration-cycle equipment is limited to a rated capacity of 65,000 BTUH (total cooling output or heating mode output). Equipment that falls into this category is subject to the minimum efficiency requirements of the National Appliance Energy Conservation Act of 1986.

6-2 Ratings and Performance Standards

Various organizations are responsible for testing and certifying the performance of primary equipment. For example, the Air Conditioning and Refrigeration Institute (ARI) produces standards for rating central, and packaged terminal, air conditioning and heat pump equipment; the Gas Appliance Manufacturers Association (GAMA) is responsible for rating fossil fuel furnaces and boilers; and the Association of Home Appliance manufacturers (AHAM) is concerned with consumer products such as room air conditioners.

After the equipment is tested in accordance with the appropriate standard, the data is compiled, formatted and published in product directories. The information presented in these directories is summarized at the top of the following page (refer to Figure 6-1).

Note that the certification data that appears in the product directories should not be used to select primary equipment or to estimate operating costs. As explained in **Manual S**, Section 8, this information is only useful for comparing the efficiency of similar types of equipment subjected to a standardized set of test-stand conditions.

6-3 Application Data

When a comfort conditioning system is designed, the equipment selection and sizing decisions must be based on comprehensive performance data. This information can be found on the engineering data sheets that are published by first-line equipment manufacturers. As explained in **Manual S**, Appendix 2, these summaries correlate delivered capacity (heating, sensible cooling, or latent cooling) with a wide range of operating conditions. For example, the cooling capacity of an air-cooled cooling unit or an air-source heat pump depends on the air-flow rate (CFM) at the indoor coil, the outdoor dry-bulb temperature, the indoor dry-bulb temperature, and the indoor wet-bulb temperature (see Figure 6-2 on page 6-4). In addition to capacity data, the system designer needs information about other issues, such as the allowable temperature rise across a furnace heat exchanger, the air-moving capabilities of the blower, and the pressure drops associated with ancillary

		Rated Performance Central Cooling and Heating Equipment						
Equipment	Comparative [1] Operating Cost Value	Efficiency Rating [2]		Cooling Performance BTUH		Heating Performance BTUH		Reference
		Cooling BTUH / W	Heating BTUH / W					
Air Conditioners	Annual Cooling	SEER		Total Capacity [3] 95 / 80 / 67 °F				R1
Air-source Heat Pumps	Annual Cooling Annual Heating	SEER	HSPF	Total Capacity [3] 95 / 80 / 67 °F		Output at 47°F Outdoor Temperature		R1
Water-source Heat Pumps	No Information	EER	COP	Total Capacity 50°F EWT	Total Capacity 70°F EWT	Output 50°F EWT	Output 70°F EWT	R2
Gas or Oil Furnace	No Value Listed		AFUE			Input Required	Output Capacity	R3
Gas or Oil Boiler	No Value Listed		AFUE			Input Required	Output Capacity	R3

References

R1) ARI Directory of Certified Products — Unitary Air-Conditioners, Unitary Air-Source Heat Pumps
R2) ARI Directory of Applied Certified Products — Unitary Water-source Heat Pumps
R3) GAMA — Consumers Directory of Certified Furnace and Boiler Efficiency Ratings

Notes

1) Op-cost values are based on a standardized scenario. They cannot be assumed to apply to a specific installation.
2) SEER and HSPF are based on a standardized scenario. They cannot be assumed to apply to a specific installation.
3) Outdoor temperature 95°F, indoor dry-bulb temperature 80°F, indoor wet-bulb temperature 67°F.
4) EWT refers to the entering water temperature.

Figure 6-1

air-side devices such as accessory filters and add-on coils (refer to **Manual D**, Section 6).

6-4 Installation Procedures

A considerable amount of the information that pertains to installation procedures and practices is generic. This knowledge has been collected in books sold by technical publishing houses and in the manuals and study courses that have been developed by trade associations such as the Refrigeration Service Engineers Society (RSES) and the National Joint Steamfitter-Pipefitter Apprentice Committee (NJS-PAC). Generic information also is included in the installation and service literature published by equipment manufacturers. These publications are useful for general training, and they are mandatory reading when they contain proprietary product information.

6-5 Safety, Combustion Air, and Venting Standards

Some agencies and associations are predominately concerned with health and safety. For example, documents and reports published by organizations such as Underwriters Laboratories (UL), the National Fire Protection Association (NFPA), and the American Gas Association (AGA) focus on manufacturing and installation practices. This guidance typically pertains to issues like minimum clearance requirements (between combustible materials and equipment cabinets or vent pipes), combustion air requirements (as discussed in this manual in Section 2), and electrical wiring standards. Also note that comprehensive vent sizing tables can be found in the Nation Fuel Gas Code and in other documents published by the AGA and GAMA.

6-6 Codes and Regulations

The efficiency and safety standards published by various agencies and associations become mandatory requirements when they are incorporated into a building code or a utility regulation. If an adopted standard pertains to the performance of a manufactured product subject to one or more labeling laws, compliance is demonstrated by the seal of approval or certificate of compliance supplied with the product. If the adopted standard pertains to an installation practice, compliance is demonstrated by the paper work that an agent of the governing authority issues after performing the on-site inspection.

6-7 Heat Sources and Heat Sinks

The concept of a heat source or a heat sink is associated with equipment that is designed to move heat. When heat is extracted from a fluid, the name of the fluid identifies the source. For example, heat pumps are commonly identified as air-source or water-source devices. When heat is rejected to a fluid, the name of the fluid identifies the sink. For example, cooling equipment is either air-cooled or water-cooled.

Outdoor air is the predominate source/sink fluid because it is readily available and because there are no development costs associated with its use. However, from the standpoint of the demand on the electrical service (KW draw), efficiency, and operating cost, outdoor air is a less desirable fluid than water; in this regard, a closed-loop system is less efficient than an one-pass system.

- When heat is rejected to a sink, the seasonal average condensing temperature associated with water-cooled equipment is lower than the seasonal average condensing temperature associated with air-cooled equipment.

- When heat is extracted from a source, the seasonal average evaporator temperature associated with water-source equipment is warmer than the seasonal average evaporator temperature associated with air-source equipment.

- The average seasonal temperatures associated with a one-pass water system are relatively constant and approximately equal to the ground water temperature. This means that one-pass water will, on an seasonal average basis, be cooler than closed-loop water during the summer and warmer than closed-loop water during the winter.

The disadvantages associated with using water as a source/sink fluid is related to installation costs and the amount of land that is required to install the system. As indicated below, these parameters are directly affected by the configuration of the system.

- A supply well, and possibly a return well, will be required for open (one-pass) systems. (The return well requirement depends on local ground water regulations and the potential for encroaching on neighboring property rights).

- If a closed-loop system is featured, a substantial amount of pipe will have to be installed in one or more wells (vertical installations are compatible with limited acreage and rocky substrate), or buried in a trench (horizontal systems are compatible with large lots and soft soil).

Unless the prospective owner is adamant about using water as a source/sink, the costs of developing the water-side of a system should be balanced against an anticipated reduction in operating costs. This pay-back analysis requires a calculation tool (computer program) that is comprehensive, technically correct, and unbiased. The effects of utility incentives such as cash rebates, special energy use rates, and relaxed loan terms also should be considered.

6-8 Central Cooling Equipment

When cooling is required, it usually is supplied by motor-driven, compression cycle air conditioners and heat pumps. As indicated in Section 5, of this manual, cooling-only equipment and air-source heat pumps could feature an outdoor condensing unit paired with a remote air handler or a refrigerant coil located at the discharge opening of a furnace, or all the machinery could be packaged in a single cabinet. Cooling also could be provided by a water-source heat pump package. (Engine driven equipment and absorption-cycle equipment is available, but this type of equipment has not captured a significant share of the residential market.)

Motor driven equipment is manufactured in single speed, multiple speed, and variable speed configurations. Because the capacity can be adjusted to match the load, there are comfort and operating cost advantages associated with equipment that features speed control, but this type of machinery is more expensive than single speed equipment.

Applied Performance Data
The performance of motor driven, compression-cycle cooling equipment is characterized by its sensible cooling capacity, latent cooling capacity, and the corresponding power input requirement. Since these parameters are affected by the operating conditions at the evaporator and the condenser, tables must be used to correlate the input and outputs with a variety of operating temperatures and flow rates.

Air-cooled Equipment Capacity
If the equipment uses the outdoor air as a sink, the sensible and latent capacities and the power input vary with the outdoor temperature, the air-side flow rate at the evaporator (blower CFM), the dry-bulb temperature of the air entering the indoor coil, and the wet-bulb temperature of the air entering the indoor coil. These relationships are summarized by tables similar to the example provided by Figure 6-2, at the top of the next page. (The exhibits and examples presented in this manual are not generic. Each equipment manufacturer uses a proprietary format when publishing equipment performance data.)

Water-cooled Equipment Capacity
If the equipment uses a water system as a sink, the sensible and latent capacities and the power input vary with the water temperature, the water-side flow rate (pump GPM), the air-side flow rate (blower CFM), the dry-bulb temperature of the air entering the indoor coil, and the wet-bulb temperature of the air entering the indoor coil. On the next page, Figure 6-3 provides an example of performance data associated with a 50 °F entering water temperature. (This matrix was extracted from a larger table that correlates performance with entering water temperatures that range from 30 °F to 110 °F.)

Wet Bulb (°F)	Air Vol. (CFM)	Outdoor Air Temperature Entering Outdoor Coil (°F)												
		85					95					105				
		Total Cool Cap. (BTUH)	Comp. Motor Watts Input	Sensible to Total Ratio S/T			Total Cool Cap. (BTUH)	Comp. Motor Watts Input	Sensible to Total Ratio S/T			Total Cool Cap. (BTUH)	Comp. Motor Watts Input	Sensible to Total Ratio S/T		
				Dry Bulb					Dry Bulb					Dry Bulb		
				76	80	84			76	80	84			76	80	84
63	1,000	29,900	2,820	.76	.88	.97	28,300	3,000	.78	.90	.97	26,700	3,190	.80	.93	.97
	1,125	30,600	2,840	.79	.92	.97	28,900	3,020	.82	.95	.97	27,200	3,210	.84	.97	.97
	1,250	31,200	2,850	.82	.96	.97	29,400	3,040	.85	.97	.97	27,900	3,250	.88	.97	.97
67	1,000	31,800	2,870	.59	.71	.82	30,100	3,070	.61	.73	.84	28,300	3,270	.62	.75	.87
	1,125	32,400	2,890	.61	.74	.86	30,500	3,080	.63	.76	.88	28,700	3,290	.65	.78	.91
	1,250	32,800	2,900	.63	.77	.89	30,900	3,100	.65	.79	.92	29,100	3,300	.67	.82	.96

Note: All values are gross capacities and do not include indoor blower coil motor heat deduction.

Figure 6-2

W-36		Cooling Capacity Data			
		1,200 CFM			
EWT °F	GPM	Entering db / wb	Total Capacity	Sensible Capacity	Input Kw
50	4.5	75/63	38.7	27.6	2.67
		80/67	42.1	28.7	2.72
	7.0	75/63	39.9	28.1	2.57
		80/67	43.4	29.3	2.63

Figure 6-3

Blower Data

Blower data is very important because it provides the basis for the duct sizing calculations. Normally, this information is presented in a tabular form (Figure 6-4), but the same data could be presented graphically. Also note that when a coil and a filter are packaged with the blower, the data-set is usually adjusted for the internal resistance created by these devices, but this is not always true. Refer to **Manual D** for a comprehensive discussion of this subject.

H-80 Blower Performance					
Air Flow (CFM) versus Fan Speed					
Fan Speed	External Resistance (IWC)				
	0.20	0.30	0.40	0.50	0.60
Low	770	770	760	735	690
Med-low	935	925	905	670	830
Med-high	1,200	1,180	1,145	1,105	1,045
High	1,360	1,315	1,265	1,195	1,125

Figure 6-4

Air-side Devices

If a cooling coil is added to a furnace, the available external pressure produced by the blower has to be reduced by an amount equal to the pressure drop across the accessory device. This information, which must be provided by the company that manufacturers the coil, is usually summarized by a table (Figure 6-5). Similar tables correlate the pressure drop across other types of add-on devices, such as an electronic filter.

DX Coil Resistance (IWC)		
CFM	Dry	Wet
1000	0.11	0.18
1200	0.15	0.26
1400	0.22	0.35

Figure 6-5

Equipment Selection and Sizing Limits

The performance of residential cooling equipment should be matched to the **Manual J** loads and to the operating conditions that apply to the design. A comprehensive discussion of this subject, as it pertains to air-cooled and water-cooled equipment, is provided in **Manual S**. This reference emphasizes the following points:

- Cooling-only equipment (air-source or water-source) should be sized so that the total cooling capacity does not exceed the total cooling load by more than 15 percent.

- If air- or water-source heat pump equipment is installed in a mild or hot climate, the total sensible and latent cooling capacity should not exceed the total cooling load by more than 15 percent.

- If air- or water-source heat pump equipment is installed in a cold climate (where heating costs are a primary concern), the total sensible and latent cooling capacity can exceed the total cooling load by as much as 25 percent. (This allows the designer to place more emphasis on refrigeration-cycle heating).

- The minimum heating season balance point is defined by the 25 percent oversizing limit on cooling capacity. This oversizing limit should not be exceeded in an attempt to satisfy an arbitrary balance point requirement. (A maximum allowable balance point mandate usually is associated with a desire to reduce the generating load on a winter-peaking utility.)

- Equipment selection decisions should be based on the applied performance data published by first-line equipment manufacturers.

- The design value for the operating temperature of an air-cooled condenser will depend on the local outdoor design temperature and the location of the equipment. (The air temperature near roofs and sun-soaked masonry is warmer than the ambient outdoor temperature.)

- The design value for the operating temperature associated with a water-cooled condenser will depend on the local ground water temperature (one-pass system); or on the local climate, soil conditions, and pipe-loop design (closed-loop system).

- The condition of the air entering the evaporator will depend on the temperature and humidity in the conditioned space, the amount of outdoor air used for ventilation (ducted to the return-side of the coil), and, if return ducts are installed in an unconditioned space, the size of the return-side leakage and conduction losses.

Operating and Safety Controls

Examples of the controls and devices that operate and protect the cooling equipment include refrigerant-side pressure switches, motor overload protection devices, anti-short-cycle relays, motor starting kits, crankcase heaters, low ambient kits, fan interlocks, speed control hardware, and a variety of thermostat options. Depending on the product, some or all of these devices are provided as standard equipment, but if an important device is not included with the basic package, it should be specified as an accessory when the equipment is ordered.

Efficiency Ratings

A seasonal energy efficiency rating (SEER) is assigned to air cooled equipment and an energy efficiency ratio (EER) is assigned to water-source heat pump equipment. This product-specific performance data can be found in the ARI directories or in engineering literature published by equipment manufacturers. This information can be used to compare the efficiency of similar types of equipment on a relative basis, but it should not be used to estimate the amount of energy that will be required for a specific home.

SEER

The SEER value is an average seasonal efficiency rating, expressed as BTUH of cooling delivered per Watt of electrical power consumed. The calculation procedure used to generate an SEER number considers the energy consumed by the condensing unit, the indoor blower, the crankcase heater, and the controls. It also accounts for the inefficiency caused by the start-stop cycles. However, this efficiency index does not apply to a specific installation because the calculation is based on assumptions regarding the relationship between the design cooling load and the installed cooling capacity, the distribution of the bin cooling hours (see **Manual J**, Table A4-1), the effects of internal loads and solar gains, and the condition of the air entering the indoor coil.

- The published SEER rating is based on an assumption that the installed capacity is equal to the design load. A correction is required for installations that involve other load-to-capacity ratios.

- The local weather patterns, as summarized by a matrix of bin cooling hours, affect the SEER value. (According to documents published by the Department of Energy, the range of excursion should be within 10 percent of the published value.)

- The published SEER value is based on a load-line model drawn from the design cooling load condition to a 65°F no-load condition. This attempt to compensate for the effects of internal loads and solar gains should not be assumed to apply to the internal load schedules and solar load patterns associated with a particular home. (Discounting the behavior of the occupants, solar loads are affected by time of year, window shading devices, cloud cover, and altitude).

- The published SEER value is based on wet-coil tests that are conducted with an 80 dry bulb, 67 wet bulb condition at the face of the indoor coil. Since the performance of the equipment is sensitive to these parameters, and since this operating condition is not generic, a correction is required for the entering condition associated with a specific installation. (The condition of the entering air depends on local weather conditions; the amount of envelope leakage; the amount of outdoor air used for ventilation; the rate of internal moisture generation; the amount of excess cooling capacity; and the location, insulation, and tightness of the return duct system.)

EER

An EER rating could apply to air- or water-source equipment, but it is commonly used to indicate the efficiency of water-source equipment. This index is a BTUH per Watt rating that represents the efficiency associated with continuous, full-load operation. Since part-load performance is not considered, an EER value would not represent the seasonal efficiency of a piece of equipment. Actually, two EER values can be found in a directory of certified products, one for 50°F entering water temperature and one for 70°F entering water temperature (Figure 6-1).

6-9 Air-to-Air Heat Pumps

Air-to-air heat pumps are motor-driven, compression cycle devices that are predominately installed in a split configuration, but single-package equipment is available. If a split system is featured, the heat pump equipment could be the sole heat source or it could be combined with a fossil fuel furnace. If it is the sole source of heat, the refrigeration-cycle equipment usually will be supplemented by accessory electric resistance heating coils that are activated by the second stage of the indoor thermostat. If a furnace is paired with the heat pump, the furnace will provide the second-stage heat, so the electric heating coils will not be required. (Second-stage heat is not required if the refrigeration cycle equipment can satisfy the design heating load. This scenario is compatible with sub-tropical climates such as Miami Florida.)

Most air-source heat pumps operate at one speed, but speed-adjustable equipment is gaining favor because the capacity can be matched to the load on a seasonal basis. This capability translates into more comfort, less resistance heating, and reduced operating cost. The primary disadvantage of this technology is that it increases the installation cost.

Applied Performance Data — Cooling
When an air-to-air heat pump operates in the cooling mode, the performance is characterized by its sensible capacity, its latent capacity, and the corresponding power input requirement. In this mode of operation, the air-source heat pump is no different than a cooling-only unit. This means that the tables that are used to document the performance of an air-to-air heat pump and the tables that summarize the performance of cooling-only equipment are similar (refer to Figure 6-2 on page 6-4).

Applied Performance Data — Heating
When operating in the heating mode without interruption, the capacity of an air-to-air heat pump varies with outdoor temperature, increasing as the outdoor temperature rises. This behavior can be summarized by a table or graph that correlates heating output with an appropriate range of outdoor air temperatures (Figure 6-6).

The heating capacity of an air-source unit also depends on the action of the defrost cycle, which causes the equipment to operate in the cooling mode for a few minutes per hour. This means that the effective heating capacity will be less than the heating capacity associated with continuous, single-mode operation. This behavior is characterized by a dip in the performance curve that is commonly referred to as the "defrost knee" (Figure 6-7).

Usually, the manufacturer's performance data includes an allowance for the effect of the defrost cycle, but this is not a

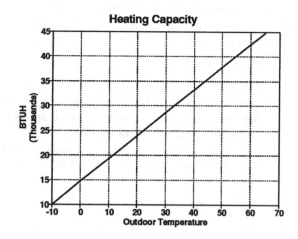

Figure 6-6

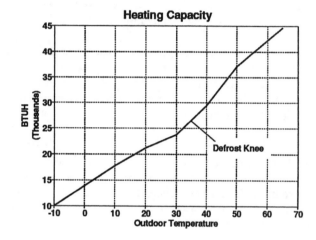

Figure 6-7

universal practice. When using performance data, always look for text or a footnote that confirms that the heating capacity data has been adjusted for the defrost penalty, or see if there is a defrost knee associated with a graphical representation of the tabulated data.

Air-Side Performance Data
As explained in the discussion of cooling equipment, the manufacturer's blower table (Figure 6-4) is always required for the duct sizing calculations. The manufacturer also is obligated to provide pressure drop data for any standard or accessory device not yet installed when the blower performance test was performed.

Defrost Cycle
During heating-mode operation, the outdoor coil is colder than the outdoor air, and depending on the amount of moisture in the outdoor air, it may be colder than the dewpoint of the

outdoor air. When this condition occurs, moisture will accumulate on the outdoor coil in the form of frost or ice (when the outdoor temperature is below 45°F). This cladding must be stripped from the coil because it degrades the performance of the equipment. Removal is normally accomplished by operating the heat pump in the cooling mode for a few minutes. In this configuration, hot gas will flow from the discharge-side of the compressor through the outdoor coil, melting the offending layer of material.

Defrost Penalties
The defrost penalty documented in an equipment manufacturer's publication corresponds to an ARI test chamber scenario that produces a mild frosting condition. The actual, on-site defrost penalty could be more or less severe than the ARI test chamber penalty, depending on the local climate and the type of defrost control. In this regard, a wet winter climate or a short timed-defrost cycle will produce a larger-than-tested defrost penalty. Conversely, a small or negligible defrost penalty is associated with a site in an arid climate. This behavior is demonstrated by Figure 6-8, which relates the defrost load to the temperature and humidity of the outdoor air. Refer to **Manual H**, Section 3 for more information about this subject.

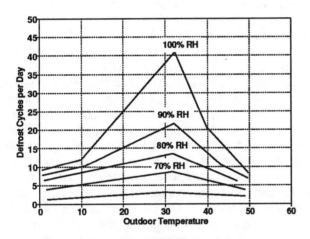

Figure 6-8

Sizing Limits
When an air-source heat pump provides heating and cooling, the size of the equipment is dictated by the cooling load. As explained in section 6-8, excess cooling capacity is limited to 15 percent or 25 percent of the total (sensible plus latent) **Manual J** cooling load, depending on the nature of the local climate. This guideline is justified by the desire to control the indoor humidity during part-load operating conditions (refer to section 1-6). However, if the local climate is characteristically dry, humidity control is not an issue. In this case, the amount of excess cooling capacity might exceed the 25 percent limit, providing that the extra capacity can be justified by

an economic analysis. However, the equipment is still subject to "reasonable size" limitations, as far as short operating cycles are concerned. (Refer to **Manual S**, for more information about optimizing equipment capacity for heating-only applications.)

Balance Point
The "balance point" refers to the outdoor temperature associated with an equivalence between the heating load and the compression-cycle heating capacity of the heat pump. This balancing act is possible because the structure and the heat pump have opposite reactions to a change in the outdoor temperature, as indicated by Figure 6-9.

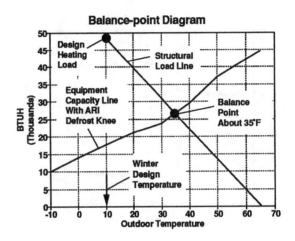

Figure 6-9

Actually, there are many possible balance points associated with an installation. Figure 6-9 represents a special case that is familiar to system designers. When this diagram is drawn (see **Manual S**, Section 4), a standard methodology is used to account for the effects of solar gains and internal loads (note the 65°F change-over point associated with the structural load line), and there is an assumption that the installed performance will mimic the performance promised by the published engineering data. In practice, this scenario does not occur with significant frequency. Solar gains and internal loads may be larger or smaller than assumed, and undocumented effects such as the actual, on-site frosting potential, a dirty filter, or an incorrect refrigerant charge will cause a reduction in the delivered heating capacity.

In any case, the design value for the balance point is the by-product of a procedure that uses the cooling load to size the heat pump equipment. This means that the designer's ability to control the balance point is bracketed by the equipment sizes that satisfy the zero to 25 percent oversizing rule. In other words, a small cooling load, relative to the size of the heating load, will most likely translate into a balance point that is relatively high, probably between 35°F and 40°F. If this is not acceptable, the heating season efficiency of the structure should be improved (basement wall insulation, for example), or fossil-fuel-heating/electric-cooling equipment can be substituted for the heat pump. Refer to **Manual S**, Section 4 for more information about balance point optimization.

Sizing Supplemental Heating Coils

The output capacity of the second-stage heating coils must equal the difference between the design heating load and the design-temperature capacity of the heat pump. Since the balance point diagram demonstrates this relationship, it can be used to determine how much supplemental heating capacity will be required for a specific application (Figure 6-10).

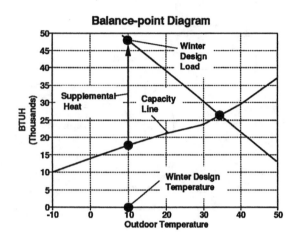

Figure 6-10

Installation of an excessive amount of supplemental capacity is not recommended because an oversized bank of resistance heating coils will degrade comfort and cause a marginal increase in the cost of operation. Also note that when a substantial amount of supplemental heat is justified by a balance point diagram, comfort will be enhanced if heating coils are activated in stages. (Refer to **Manual S**, Appendix 7 for more information about the effect of an excessive amount of supplemental heating capacity.)

Emergency Heat

Emergency heat is the total amount of resistance-coil heat that can be activated if the compressor fails. This heat can be provided by the supplemental heating coils plus a reserve bank of coils that can be activated only by a manual switch. Normally, the total capacity of both sets of coils is equal to, or slightly greater than, the design heating load, but a code or regulation may specify otherwise. Figure 6-11 shows the relationship between emergency heat and supplemental heat.

Auxiliary Heat

The term "auxiliary heat" is often used incorrectly to describe the heating coils that are activated by the second stage of the

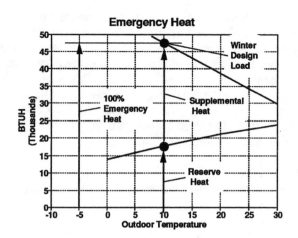

Figure 6-11

indoor thermostat. A more precise definition can be found in **Manual S**, which uses this term in reference to the total amount of resistance-coil heat installed in the heat pump package. As indicated above, some of the auxiliary heat is used to supplement the refrigeration cycle machinery during cold weather; and when the occasion arises, the reserve is used to compensate for malfunctioning machinery.

Operating and Safety Controls

The controls and devices that operate and protect air-source heat pump equipment include refrigerant-side pressure switches, motor overload protection devices, anti-short-cycle relays, motor starting kits, crankcase heaters, low ambient kits, fan interlocks, and speed control hardware. (Similar devices are supplied with cooling-only equipment.) In addition, the manufacturer will usually provide one or more defrost control options (time-temperature and demand), a suction-line accumulator, one or more outdoor thermostat options, and a line of indoor thermostats designed to control all modes of operation. As noted, when an important device is not included as standard equipment, it should be a specified accessory when the equipment is purchased from the manufacturer. (Refer to **Manual H** for more information about heat pump operating and safety controls.)

Efficiency Ratings

Air-source heat pumps have two standard efficiency ratings. When operating in the cooling mode, efficiency is characterized by the SEER value discussed in Section 6-8. When operating in the heating mode, a heating season performance factor (HSPF) is used to indicate the equipment's relative efficiency. Both of these indices are listed in the ARI Directory of Certified Products and in the engineering literature published by equipment manufacturers. (They both can be used to compare the seasonal efficiency of similar types of equipment on a relative basis, but neither should be used to estimate the amount of energy that will be required for a specific home.)

HSPF

The HSPF value is an average seasonal efficiency rating, expressed as BTUH of heating delivered per Watt of electrical power consumed. The calculation procedure used to generate an HSPF value considers the energy required for the condensing unit, the supplemental coils, the indoor blower, the crankcase heater, and the controls; and it includes a cycling penalty and a defrost penalty. However, this efficiency index does not apply to a specific installation because the calculation is based on assumptions regarding the relationship between the design heating load and the installed refrigeration-cycle heating capacity, the distribution of the bin heating hours, the effect of internal loads and solar gains, the defrost penalty, the amount of resistance heat energized during the defrost cycle, and the air temperature at the entrance of the indoor coil.

• The published HSPF rating is based on the assumption that the design heating load is equal to the heating capacity of the heat pump when the outdoor air temperature is equal to 47°F. This assumption produces a balance point diagram similar to Figure 6-12. Note that this diagram indicates that a minimum amount of supplemental heat will be required, which has the effect of maximizing the HSPF rating. (This load-capacity scenario does not apply to most of the homes located in cold climates.)

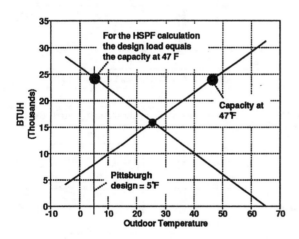

Figure 6-12

• Figure 6-13 provides a balance point diagram that represents a thermally efficient home located in a zone-4 climate. (2000 square feet of conditioned space, double pane windows, R-19 walls, R-30 ceiling, insulated slab edge, reasonable tight construction, and heat pump size compatible with the cooling load.) When Figure 6-13 is compared with Figure 6-12, it is obvious that more supplemental heat will be required for an actual zone-4 home (compare triangle A, B, C with triangle A, D, E). It follows that the HSPF rating associated with Figure 6-13 will be smaller than the HSPF rating produced by Figure 6-12.

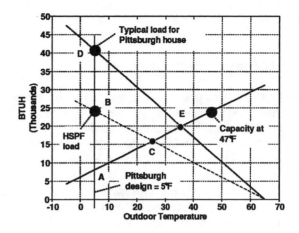

Figure 6-13

• The published HSPF rating is based on a climate model that simulates the weather in Pittsburgh, Pennsylvania. Everything else being equal, different rating values would be expected to apply to other climates. In the case of air-source heat pumps, this effect cannot be ignored, so six different efficiency ratings have been developed to accommodate a variety of climate zones; but the zone-4 index (Pittsburgh) is the only value that commonly is published.

• The published HSPF value is based on a load-line model that is drawn from the design heating load condition to a 65°F no-load condition. This attempt to compensate for the effects of internal loads and solar gains is commendable, but it cannot be assumed to be representative of the internal and solar load schedules associated with a particular home. (Discounting the behavior of the occupants, solar loads are affected by time-of-year, window shading devices, cloud cover, and altitude).

• As noted above, the defrost penalty associated with the certification test is based on a scenario that creates a mild frosting condition. This means that an HSPF correction is required when the actual, frosting potential is substantially larger or smaller than the assumed frosting potential.

• If the amount of resistance heat energized during the defrost cycle is greater than the amount of heat required to neutralize the associated cooling effect, heat will be added to the structure at a COP of 1.0. This means that when the heating mode is restored, an equivalent amount of heat will not be required. This has an a small effect on the HSPF because this heat could have been delivered by the heat pump at a higher COP.

• The performance of an air-source heat pump is slightly affected by the temperature of the air entering the indoor coil. (The temperature of the entering air depends on local weather condition; the amount of outdoor air used for ventilation; and the location, insulation, and tightness of the return duct system.)

6-10 Water-to-Air Heat Pumps

A water-to-air heat pump is a motor-driven, compression cycle device that features a water-to-refrigerant heat exchanger. This piece of heat transfer hardware provides the interface between the refrigeration-cycle equipment and a storage reservoir, which could be provided by an open, well water system, or a closed, earth-coupled, water-filled, piping system. Because this heat exchanger does not have to be located in the storage reservoir, all the refrigeration-side and air-side devices can be packaged in one cabinet. This means that two pipes (a supply and a return) will provide the interface between the comfort conditioning equipment and the water system.

Normally, a water-source heat pump will be the sole source of heat (this equipment is not usually paired with a fossil fuel furnace), which means that the refrigeration-cycle equipment may have to be supplemented by electric resistance heating coils that are activated by the second stage of the indoor thermostat. (Second stage heat is not required if the refrigeration cycle equipment can satisfy the design heating load when the winter design conditions are in effect. This scenario is compatible with home sites located in mild and hot climates.)

Single-speed, water-source equipment characteristically is efficient, but speed control is still desirable because a close match between compressor capacity and the seasonal load will translate into more comfort, less demand for second-stage heat, and lower operating cost. Of course, these benefits must justify the additional expense of installing a more sophisticated product.

Applied Performance Data — Cooling
When a water-source heat pump operates in the cooling mode, the performance is characterized by its sensible capacity, latent capacity, and the corresponding power input requirement. These parameters are, in turn, affected by the entering water temperature, the water-side flow rate, the air-side flow rate, and the condition of the air at the entrance to the air-side coil. These concepts have previously been introduced in section 6-8 and summarized by Figure 6-3.

Latent Capacity
In regard to dehumidification capability, water-source equipment could have significantly more latent capacity than a similarly sized air-source package, or about the same amount of latent capacity as an air-source unit, depending on the entering water temperature. If the entering water temperature is relatively low, between 30°F and 70°F (typical of a one-pass, well water system), the performance of the water-source equipment is characterized by a low coil sensible heat ratio. But, if the entering water temperature is comparable to the

summer outdoor design temperature, 80°F and warmer (a characteristic of a closed loop system), the coil sensible heat ratio associated with the water-source equipment will be similar to an air-source coil sensible heat ratio. (A comprehensive discussion of this subject can be found in **Manual S**, Section 3.)

Applied Performance Data — Heating
When operating in the heating mode without interruption, the heating capacity of a water-to-air heat pump is very sensitive to the entering water temperature, much less sensitive to the water-side flow rate, and marginally sensitive to the air temperature at the entrance to the indoor coil. Manufacturers typically summarize these relationships by using tables similar to Figure 6-14.

W-41		Heating Capacity Data		
		1,300 CFM		
EWT °F	GPM	Entering Dry-bulb	Heating Capacity	Input KW
30°F	4.5	60	26.8	2.45
		70	25.9	2.58
		80	24.9	2.71
	7.0	60	28.0	2.52
		70	27.0	2.66
		80	26.0	2.79
	9.0	60	29.3	2.60
		70	28.3	2.74
		80	27.2	2.88
50°F	Actual table would include data for 50°F EWT			
70°F	Actual table would include data for 70°F EWT			
90°F	4.5	60	53.2	3.66
		70	51.9	3.85
		80	50.7	4.04
	7.0	60	55.3	3.77
		70	54.0	3.97
		80	52.6	4.17
	9.0	60	57.3	3.85
		70	55.9	4.05
		80	54.5	4.25

Figure 6-14

Optimum Water-Side Flow Rate
The optimum flow rate associated with the water-side of the system depends on the type of water system. If a one-pass, well water system is involved, the lowest flow rate sanctioned by the manufacturer's engineering data will suffice, because there is only a small performance advantage associated with an increased flow rate, and because water use and pumping power should be minimized as a matter of policy. But, if a closed, earth-coupled piping system is featured, the water-

side flow rate must be large enough to insure turbulent flow throughout the piping system. This means that the HVAC system designer and the water-loop system designer must agree on a water-side flow rate that is compatible with the equipment performance requirement and the turbulent flow requirement associated with the water-side of the system.

Air-Side Performance Data

As explained in the discussion of cooling equipment, the manufacturer's blower table (Figure 6-4) is always required for the duct sizing calculations. The manufacturer also is obligated to provide pressure drop data for any standard or accessory device that was not installed when the blower performance test was performed.

Defrost Cycle

There is no defrost cycle associated with water-source equipment, but this does not mean that the temperature of the water flowing to the inlet of the water-to-refrigerant heat exchanger will never be below freezing. If an earth-coupled, water-loop system is installed in a colder climate (occasional below-zero temperatures), the piping circuits will have to be charged with a solution that has an appropriate mixture of water and anti-freeze solute.

Water-Side Design and Installation

The water-side of a water-source heat pump system could be designed and installed by the HVAC contractor, a well drilling company, or an excavation/piping company. If a one-pass, well water system is required, the HVAC contractor only needs to specify a minimum water-side flow rate and then hook up to the water system provided by the sub-contractor. If a closed, earth coupled piping loop is installed, the HVAC contractor and the pipe system designer will have to collaborate as they work through the design rubric. More information about the details of designing and installing water systems can be found in the **Geothermal Heat Pump Training and Certification Program** training manual and the **Earth Coupled Heat Pump Water-Loop Design** document, as published by ACCA. Also refer to ACCA **Manual H** and to the **Closed-Loop/Ground-Source Heat Pump Systems Installation Guide** that is published by the National Rural Electric Cooperative Association, Washington DC.

Sizing Limits

When a water-source heat pump provides heating and cooling, the size of the equipment is dictated by the cooling load. As explained in section 6-8, excess cooling capacity is limited to 15 percent or 25 percent of the total (sensible plus latent) **Manual J** cooling load, depending on the nature of the local climate. This guideline is justified by the desire to control the indoor humidity during part-load operating conditions (refer to section 1-6). However, if the local climate characteristically is dry, humidity control is not an issue. In this case, the amount of excess cooling capacity might exceed the 25 percent limit, providing that the extra capacity can be justified by an economic analysis. However, the equipment is still subject to "reasonable size" limitations, as far as short operating cycles

are concerned. (Refer to **Manual S**, for more information about optimizing equipment capacity for heating-only applications.)

Balance Point

The term "balance point" refers to the outdoor temperature associated with an equivalence between the heating load and the compression-cycle heating capacity of the heat pump. This point of equilibrium occurs even though the outdoor temperature affects only the performance of the structure. (The performance of the heat pump is dictated by the entering water temperature, which is either indifferent to, or only slightly affected by, the outdoor temperature, depending on the type of water system — which could be open or closed.) An example of a water-source heat pump balance point diagram is provided by Figure 6-15.

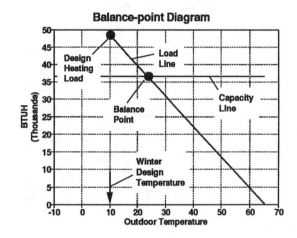

Figure 6-15

With any type of heat pump system, there are many possible balance points associated with the installation. Figure 6-15 represents the balance point diagram used by system designers. When this diagram is drawn (see **Manual S**, Section 5), a standard methodology is used to account for the effects of solar gains and internal loads (note the 65°F change-over point associated with the envelope load line), and it is assumed that the installed performance will mimic the performance promised by the published engineering data. In practice, this scenario does not occur with significant frequency. Solar gains and internal loads may be larger or smaller than assumed, and undocumented effects such as a variation in the entering water temperature, fouled heat exchanger tubes, a dirty filter, or an incorrect refrigerant charge will cause a reduction in the delivered heating capacity.

In any case, the design value for the balance point is the by-product of a procedure that uses the cooling load to size the heat pump equipment. This means that the designer's ability to control the balance point is bracketed by the equipment sizes that satisfy the zero to 25 percent oversizing rule. In other words, a small cooling load, relative to the size of the heating load, could translate into a balance point that is higher than desired. If a lower balance point is desired, the heating season efficiency of the structure should be improved (basement wall insulation, for example). Refer to **Manual S**, Section 5 for more information about balance point optimization.

Sizing Supplemental Heating Coils

The balance point diagram demonstrates how much supplemental heating capacity will be required for a specific application. As indicated by Figure 6-16, the output capacity of the second-stage heating coils is equal to the difference between the design heating load and the design-temperature capacity of the heat pump.

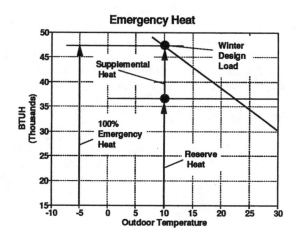

Figure 6-17

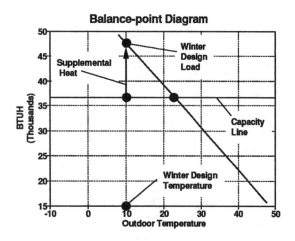

Figure 6-16

Normally, the amount of resistance coil heating capacity associated with a water-source heat pump will be significantly less than the second-stage capacity required for an equivalent air-source package. In either case, installing of an excessive amount of supplemental capacity is not recommended because an oversized bank of resistance heating coils will degrade comfort and cause a marginal increase in the cost of operation. (Refer to **Manual S**, Appendix 7 for more information about the effect of an excessive amount of supplemental heating capacity.)

Emergency Heat

Emergency heat is the total amount of electric resistance heat that can be activated if the compressor fails. This heat can be provided by the bank of supplemental heating coils plus a reserve bank of coils that can be activated by a manual switch. Normally, the total capacity of both banks of coils is equal to, or slightly greater than, the design heating load, but a building code or regulation may specify otherwise. Figure 6-17 shows the relationship between emergency heat and supplemental heat.

Auxiliary Heat

As explained in Section 6-9, auxiliary heat is the total amount of resistance-coil heat installed in the heat pump package. As indicated by Figure 6-16, some of the auxiliary heat is used to supplement the refrigeration cycle machinery during cold weather. When the occasion arises, the reserve is used to compensate for malfunctioning machinery.

Operating and Safety Controls

The controls and devices used for operating and protecting water-source heat pump equipment includes refrigerant-side pressure switches, motor overload protection, anti-short-cycle relays, motor starting kits, freezestats, water-side solenoids, fan interlocks, and speed control hardware. In addition, the manufacturer may offer an outdoor thermostat option and will provide a line of indoor thermostats designed to control all modes of operation. When an important device is not included as standard equipment, it should be a specified accessory when the equipment is purchased from the manufacturer. (Refer to **Manual H** and the **Geothermal Heat Pump Training Certification Program** manual for more information about water-source heat pump controls.)

Efficiency Ratings

Water-source heat pumps have two standard efficiency ratings. When operating in the cooling mode, efficiency is characterized by the two EER values discussed in Section 6-8. When operating in the heating mode, a coefficient of performance value (COP) is used to indicate efficiency. This parameter is a dimensionless number that represents the ratio of output energy to input energy. Also note that the COP rating is associated with continuous full-load operation, so it does not indicate the seasonal efficiency of the equipment. Actually, two COP values can be found in the ARI Directory of Certified Products; one indicates the efficiency associated with a 50°F entering water temperature and the other indicates the efficiency associated with a 70°F entering water temperature (see Figure 6-1).

6-11 Air-to-Water or Water-to-Water Heat Pumps

Air-to-water and water-to-water heat pump equipment is available, but it is not commonly used for residential applications. If installed, it would be used in conjunction with terminal devices such as a hot water coil (located in an air handler), fin-tube convectors, or fan-powered convectors. Refer to **Manual H** for more information about this type of equipment.

6-12 Multi-Zone Equipment

The are two basic types of residential multi-zone systems. The variable volume approach combines a central air handler with an electronic control system and a set of automated throttling dampers; and the multiple split-coil concept features a set of indoor fan-coil units that are served by a single outdoor unit containing the balance of the refrigeration-cycle devices. In the first case, the primary energy conversion equipment is similar to the products used for a single-zone applications. In the second case, the equipment package is specifically designed for zone control. Simple examples of both types of installations are provided by Figures 18 and 19.

Variable Air Volume System
- Air flow is controlled by zone dampers that are operated by a local thermostat.
- The central equipment is controlled by a microprocessor that monitors the zone thermostats.
- The blower CFM could be constant, but a bypass duct (with modulating damper) will be required to maintain the air flow through the equipment.
- The blower CFM could vary if the fan and compressor speed can be automatically synchronized with the load.

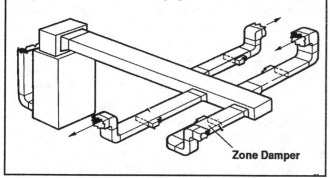

Zone Damper

Figure 6-18

Multi-Zone Load Calculations
The load calculation procedure that applies to zoned systems is different than the traditional (single zone) computation. In this case, the designer must be concerned with peak room loads, peak zone loads, and the effect of diversity as it applies to the maximum load on the central equipment. Refer to **Manual J**, Appendix A2 for more information on this subject.

Ductless Split-coil System
- Manufacturers offer cooling-only equipment and heat pump systems.
- Two to five indoor coils can be connected to one outdoor unit. (The maximum coil split depends on the product.)
- A duct system is not required when air conditioning is added to a home that has a baseboard or radiant heating system.
- Indoor coils are controlled on a zone-by-zone basis by room or zone thermostats.
- The central equipment is controlled by a microprocessor that monitors the zone thermostats and the refrigerant flows.

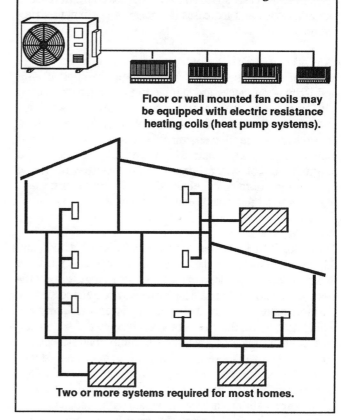

Floor or wall mounted fan coils may be equipped with electric resistance heating coils (heat pump systems).

Two or more systems required for most homes.

Figure 6-19

Equipment Selection
The procedures that are used to select multi-zone equipment are the same as the procedures that are used to select single-zone equipment. In either case, the designer must use manufacturer's application data to demonstrate that a particular equipment package will satisfy the design loads (sensible cooling, latent cooling, or heating) when the equipment is subjected to a specific set of operating conditions. These procedures are documented in **Manual S**.

Air-Side Design
The procedure used to design a variable volume duct system is not identical to the traditional, constant volume duct sizing procedure. The difference between the two procedures is associated with the design flow rate values. Variable volume

duct runs are sized to accommodate the maximum flow rates that would be experienced during any operating condition. Refer to **Manual D** for instruction on this matter.

Speed Control
If a variable volume system is equipped with adjustable speed machinery, the blower and compressor capacity can be matched to the load. This is a very desirable feature, because it usually eliminates the need for a bypass duct. (**Manual D** discusses this subject in more detail.) If a multiple split-coil system is equipped with speed control, the refrigerant flow rates can be matched to the load associated with the indoor coils. In either case, speed control enhances comfort, reduces operating cost, and increases the range of equipment-sizing options.

Operating and Safety Controls
Many of the operating and safety controls supplied with multi-zone equipment are similar to the devices packaged with single-zone equipment, but some controls may be proprietary. In the case of variable volume systems, special controls are required to coordinate the operation of the zone thermostats, volume dampers, and central equipment. If a multiple split-coil system is installed, the package will include controls that coordinate the operation of the zone thermostats, indoor fan-coil units, and central equipment.

Efficiency Ratings
Traditional efficiency descriptors such as SEER, HSPF, and AFUE are assigned to the multi-zone equipment, but since these indices were developed to valuate single-zone efficiency, they do not indicate the efficiency of a zoned system. If a more representative set of descriptors were available, they would account for the effects of load diversity and set-up-setback control strategies. On this basis, multi-zone systems are 10 to 40 percent more efficient than singe-zone systems, depending on the control strategy. Also note that multi-zone systems provide an opportunity to reduce the total design load if comfort conditions are synchronized with occupancy on a room-by-room basis. (Refer to **Manual J**, for more information about the reduction in installed capacity associated with a comfort on-demand sizing strategy.)

6-13 Dual-Fuel Heating Equipment

Dual fuel heating systems are commonly created by pairing an air-source heat pump with a fossil fuel furnace. In this case, the two devices are independently manufactured and each piece of equipment is sized by using the guidelines that apply to a single-fuel scenario. (The heat pump is sized to satisfy the design cooling load and the furnace is sized to satisfy the design heating load.)

Hybrid, single-package equipment also is available (see Section 5-3). In this case, the compression cycle components and the fossil fuel devices operate as an integrated system. This arrangement may require some electric resistance heat in the indoor air handler (depending on the size of the heating and cooling loads), and the equipment sizing guidelines are product-specific recommendations provided by the manufacturer.

Dual fuel systems normally are installed when the cost of the fossil fuel is expensive compared to the cost of electricity. This scenario is characterized by a low economic balance point (30°F or less), which is defined as the outdoor temperature associated with equal operating costs for either fuel. Conversely, a dual fuel system is not recommended if the economic balance point is relatively high. For an example, Figure 6-20 shows an economic balance point diagram that would discourage the use of a dual-fuel system.

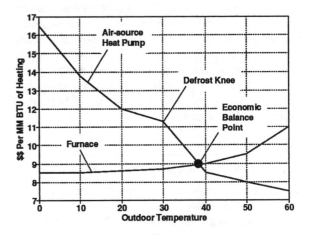

Figure 6-20

It is important to understand that a low economic balance point, by itself, only indicates that the total fuel bill can be reduced. It does not indicate whether this reduction will be large enough to justify the additional expense of installing a two-fuel system. (A comprehensive bin-hour calculation or hourly simulation is required to answer this question, and the calculations should consider any incentives offered by the local utility.) Also note that the thermal balance point concept still applies. This means that when the outdoor temperature is below the thermal balance point, but above the economic balance point, the furnace will have to provide the supplemental heat. (Manufacturers sell control packages that perform this function.)

The details associated with sizing dual-fuel equipment are documented in **Manual S**, Section 6. Additional information can be found in **Manual H**, Section 2.

6-14 Furnaces

Gas, oil, and propane furnaces are manufactured in configurations (low-boy, high-boy, up-flow, down-flow, and horizontal) that are designed to be compatible with the space to be occupied by the equipment, and with the location of supply and return duct runs (see **Manual D**, Section 2). Furnaces also are distinguished by the source of combustion air, the type of vent, the efficiency rating, and the condition of the gas in the heat exchanger (condensing or dry).

Combustion Air

Atmospheric burners take the combustion air from the surrounding space. This means that this type of equipment must either be installed in a room that has enough infiltration to satisfy the demand for combustion air, or it could be installed in a room that is equipped with structural vent openings, or it could be installed in a room that is ventilated mechanically. (See Section 2-2 for comments about combustion air and the ventilation requirements that are specified in the National Fuel Gas Code.)

Some furnaces use a fan to move air through the combustion chamber. If a fan forces air into the combustion chamber, it is referred to as a forced draft, power burner, power combustion, or pressure-fired system. If the fan exhausts the combustion chamber, it is referred to as an induced draft, induced vent, or power vent system. However, this type of equipment could still draw the combustion air from the surrounding space, which means that vent openings or mechanical ventilation will be required if infiltration cannot satisfy the demand for combustion air.

If a fan forces outside air into a sealed combustion chamber and forces the combustion gases out of the vent, it is referred to as a direct vent system. Since no combustion air is taken from inside the structure, infiltration, vent openings, or mechanical ventilation are not required for the safe operation of the equipment. (A supply of outdoor air is required for other reasons, however. Refer to Section 2.)

Venting System

Vents operate by either natural draft or a mechanical assist. If a fan is involved, the vent may operate under either a positive or negative pressure. The temperature of the combustion gases and the potential for condensation in the vent are additional factors that are used to classify a vent. Refer to Appendix 6 for a comprehensive discussion of the requirements and design procedures associated with venting gas- and oil-fired equipment.

Application Data

Manufacturer's performance data is required to select and size a furnace. Figure 6-21 shows that the published engineering data provides information about the input and output capacities associated with steady, full-load operation; the range of acceptable temperature rises across the heat exchanger; and the annual fuel utilization efficiency rating (AFUE). As explained in **Manual S**, the output capacity of the furnace must be equal to or larger than the design heating load and the temperature rise across the heat exchanger must be within the manufacturer's suggested range.

H-80 Heating Performance (BTUH) Heating (H) and Heating-Cooling (HC) Models				
Model	50	60	80	100
Input BTUH	50,000	60,000	80,000	100,000
Output BUH	39,000	47,000	62,000	78,000
AFUE	80%	80%	80%	80%
Temp. Rise	25-55	25-55	35-65	35-65

Figure 6-21

Also note that blower data and coil pressure drop data (Figures 6-4 and 6-5) are just as important as the heating capacity data because the furnace blower performance must be compatible with the duct runs and the air-side devices installed in the duct system. As explained in **Manual S**, a furnace may have adequate heating capacity and insufficient blower capacity. This situation typically occurs when a refrigerant coil is added to a furnace. (If the cooling load is large compared to the heating load, a furnace that has the correct amount of heating capacity may not be able to provide the air flow required for cooling.)

Sizing Limits

The sizing limits associated with furnaces depend on the required heating capacity and the blower performance. If possible, the furnace should not have more than 40 percent excess capacity, but this guideline can be exceeded if more blower power is required. Refer to **Manual S** for a complete discussion of this subject.

Safety and Operating Controls

The controls associated with operating and protecting furnaces include devices that monitor and operate the combustion system, the venting system, the temperature rise, and the blower motor. If an important device is not included as standard equipment, it should be a specified accessory when the furnace is purchased from the manufacturer.

Efficiency

Furnace efficiency is characterized by the Annual Fuel Utilization Efficiency (AFUE) rating. This index represents the average seasonal efficiency, expressed as the ratio of annual energy output to annual burner energy input. It does not include the electrical energy required for the blower and the controls, or the energy consumed by a standing pilot.

The calculation procedure used to generate a AFUE value includes a cycling penalty that is based on an assumption that the heating season has 5,200 heating degree days and the equipment has 70 percent excess heating capacity. According to researchers at the Brookhaven National Laboratory, cycling efficiency is not very sensitive to oversizing. This means that the published AFUE value approximately applies to other combinations of weather conditions and excess capacity. However, the range of application has not been specified.

6-15 Boilers

Fossil fuel boilers may be fired by natural gas, oil, or propane, or electric coils can be used to generate a supply of hot water. Boilers also are classified by the condition of the gas in the heat exchanger (condensing or noncondensing), the source of combustion air, the type of vent, and the efficiency rating.

Combustion Air

If a boiler is equipped with an atmospheric burner, the equipment must be installed in a room that has enough infiltration to satisfy the demand for combustion air. If infiltration cannot provide the necessary air, the room must be passively vented by an opening or duct to a source of outdoor air, or it could be vented mechanically (refer to Section 2-2).

Some designs use a fan to collect air from the surrounding space and force it through the combustion chamber. This fan could be located either upstream (forced draft) or downstream (induced draft) from the heat exchanger. But, regardless of the arrangement, vent openings or mechanical ventilation will be required if infiltration cannot satisfy the demand for air.

If direct vent equipment is installed, the combustion process will be isolated from the indoor environment. This means that infiltration, vents, or mechanical ventilation is not required for the safe operation of the equipment, but a supply of outdoor air is required for other reasons (refer to Section 2).

Venting System

Vents operate by either natural draft or a mechanical assist. If a fan is involved, the vent may operate under either a positive or negative pressure. The temperature of the combustion gases and the potential for condensation in the vent are additional factors that are used to classify a vent. Refer to Appendix 6 for additional discussion of venting requirements and information about the procedures used to size vents for gas- and oil-fired equipment.

Application Data

Manufacturer's ratings are required to select and size a boiler. Figure 6-22 shows that the published information lists the input and output capacities associated with steady, full load operation; an I-B-R net rating capacity; and a value for the annual fuel utilization efficiency.

HWB Heating Performance			
Gas-fired Models			
Model	WB-62	WB-96	WB-130
Input BTUH	62,000	96,000	130,000
Output BUH	54,000	83,000	111,000
I-B-R BTUH	47,000	72,200	96,500
AFUE	87.2	86.2	85.1

Figure 6-22

Sizing Limits

As explained in **Manual S**, the output capacity of the boiler must be equal to or larger than the design heating load for the home and the load associated with the piping losses. Or, if the I-B-R rating is used, the boiler size can be based on the **Manual J** load for the home. In this case, a piping loss estimate is not required because the I-B-R rating discounts the output capacity of the boiler by 15 percent, which according to the hydronics industry, compensates for typical piping-loss scenerios. (The piping losses are zero if all the piping is located in the heated space.)

Boilers should not be oversized to compensate for a pick-up load. Research conducted by the Hydronics Institute indicates that there is no appreciable reduction in the length of the warm-up cycle when a boiler is oversized. Even considering the capacity jumps associated with standard sizes, the output capacity should not exceed the estimated heating load by more than 40 percent.

Leaving Water Temperature

Low pressure boilers can supply water at temperatures that range from 120°F to 250°F. Within this span, the design temperature used for a particular application depends on temperatures used to select the terminal equipment. Supply temperatures below 120°F are not recommended because of problems associated with condensation in the boiler flue gas and reabsorption of air into the boiler water. Supply temperatures above 250°F require boiler operation at pressures that exceed 30 PSI.

Temperature Rise and Temperature Drop

The water temperature rise (TR) across a boiler depends on the output capacity (BTUH) and the GPM flowing through the boiler. This temperature rise can be computed by using the water-side equation.

$$TR = \frac{\text{Output BTUH}}{500 \times GPM}$$

Note that the temperature rise across a boiler is limited by the potential for causing thermal shock. Cast iron boilers can tolerate a 25°F to 30°F temperature rise. Steel boilers are less tolerant to thermal shock, so the temperature rise associated with these products should not exceed 20°F.

Also note that the temperature rise across the boiler and the temperature drops across the terminal devices are not independent of each other. If the temperature rise across the boiler is used as the design criteria, the temperature drop associated with a terminal device will be determined by the water-side equation. In this case, the *Output BTUH* is the heating capacity of the device and the *GPM* is the water flow rate through the device. The relationship between the boiler temperature rise and the terminal temperature drop also is affected by the piping arrangement.

Series Piping

Figure 6-23 shows a simple series piping arrangement. In this case 6 GPM flows through the boiler and the total load on the boiler is equal to 60,000 BTUH. This means that the water-side equation will predict a 20°F temperature rise across the boiler.

The water-side equation also can be used to calculate the temperature drop across each heating coil (piping losses are ignored, for the sake of simplicity). As indicated by Figure 6-23, the series system is characterized by a continuously decreasing temperature in the direction of flow and the total temperature drop across all the coils is equal to 20°F. A parallel piping arrangement will be required if the designer wishes to maintain a constant temperature at the entrance to each coil or to control flow rate through the coils.

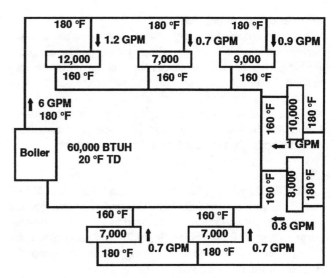

Figure 6-24

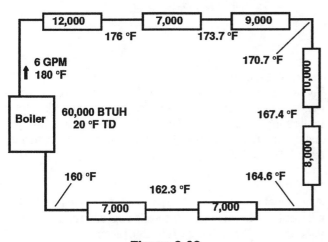

Figure 6-23

Parallel Piping

Figure 6-24 shows the same system with a parallel piping arrangement. In this case, the temperature drop across each coil must be equal to the 20°F temperature rise across the boiler, as predicted by the water-side equation. This means that the flow rates associated with the seven coils will not be equal to the 6 GPM boiler flow rate. These coil flow rates can be estimated by using the water-side equation, as indicated by Figure 6-24. Note that with this design, a constant water temperature is maintained at the entrance to each coil and the sum of the coil flow rate values is equal to the 6 GPM boiler flow rate value.

However, a parallel piping arrangement provides the option to set values arbitrarily for the coil temperature drops and the coil flow rates. Figure 6-25 shows what would happen if the designer decided to increase the temperature drops across the last two coils by 10 degrees. Note that the flow through the boiler would be reduced to about 5.6 GPM, causing the temperature rise across the boiler to increase to 21.4°F, which is slightly more than desired for a steel boiler.

Figure 6-25 also suggests that a bypass can be used to obtain a desired combination of coil temperature drops and boiler temperature rise. The calculations are not shown, but if 0.4 GPM of 180°F water is bypassed back to the boiler, the boiler temperature rise will be equal to 20°F and the coil temperature drops will be equal to 20°F (first five coils) or 30°F (last two coils), as required.

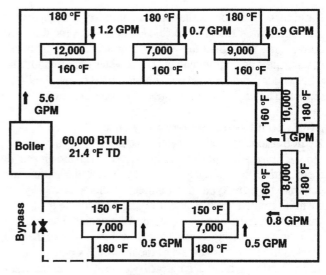

Figure 6-25

Water-Side Pressure Drop

Some hot water heating systems work by gravity, but most modern systems rely on a circulating pump. In this case, a value for the pressure drop across the water-side of the boiler is required for the pump selection calculation. Boiler manufacturers usually publish this information in their data tables.

Safety and Operating Controls

Boilers are protected by devices that monitor or manage the combustion system and the venting system, and by temperature limit switches and pressure relief valves. Other devices are used to control the water temperature and the operation of the pump. Water-side trim also includes isolation valves and a drain cock. If an important control or device is not included as standard equipment, it should be a specified accessory when the boiler is purchased from the manufacturer.

Efficiency

Boiler efficiency is characterized by the Annual Fuel Utilization Efficiency rating (AFUE). This index represents the average seasonal efficiency, expressed as the ratio of annual energy output to annual burner input. It does not consider the electrical energy required for the pump, a burner motor or the controls, or the energy consumed by a standing pilot.

The calculation procedure that is used to generate a AFUE value includes a cycling penalty, based on an assumption that the heating season has 5,200 heating degree days and the equipment has 70 excess heating capacity. But, according to the noted research, cycling efficiency is not very sensitive to oversizing. This means that the published AFUE value approximately applies to other combinations of weather conditions and excess capacity; however, the range of application has not been specified.

6-16 Hydronic Terminal Devices

Baseboard radiation is the most common type of hydronic terminal device found in homes. Other devices include wall convectors, radiators, fan-powered unit heaters, or an air handler package equipped with a hot water coil. In addition, a hot water piping circuit could be installed below the floor. (Refer to Section 5 for more about hydronic systems.)

Air Pattern Created by Buoyancy Force

By definition, convective heating equipment does not use a blower or a ducted air distribution system to create air movement within the occupied space. Typically, the air motion generated by this type of equipment is characterized by the currents created when heated air escapes from the grille of a baseboard fixture or a wall-mounted convector. These flows tend to assume a ring-shaped pattern, which moves up the wall and around the cross-section of the room. This type of circulatory motion produces comfort because it warms the floor and limits the floor-to-ceiling temperature difference to about 2°F.

Heating Capacity

The heating capacity of a convective or fan-powered device depends on the entering air temperature, the water temperature (entering or average), and the water-side flow rate. An example of baseboard convector and fan-coil data is provided by Figure 6-26. Note that the heating capacity of the baseboard convector is expressed on a per-foot basis.

Baseboard Convector Performance BTUH per Foot						
Flow Rate	Average Water Temperature °F					
	170	180	190	200	210	220
4 GPM	550	620	690	750	810	880
1 GPM	520	590	650	710	770	830
Entering air temperature = 70°F						

Fan Coil Performance BTUH			
Flow	FC-10	FC-15	FC-25
2 GPM	5,000	15,000	25,000
Entering air temperature = 70°F Entering water temperature = 200°F			

Figure 6-26

Water Temperature

The terminal device can be sized for a wide range of water temperatures. Depending on the design concept with respect to the supply water temperature, the boiler temperature rise, and the circuit temperature drop, this temperature could vary from 120°F to 250°F.

Flow Rate

A terminal device can be sized for any flow rate that is compatible with the manufacturer's data and the boiler temperature rise. Typically, the flow rate ranges between 1 GPM and 4 GPM.

Capacity, Flow Rate, and Temperature Drop

The heating capacity, flow rate, and water-side temperature drop are interdependent. If two values are known, the third can be calculated by using the water-side equation.

Water-Side Pressure Drop

A value for the water-side pressure drop will be required for the pump selection calculations. Manufacturers usually publish this information in their data tables. (Refer to Appendix 5 for more information about piping system design.)

Air-Side Pressure Drop

If a fan-coil device is shipped as a single package, information about the air-side pressure drop across the coil is not required. However, pressure drop information is required when a hot water coil is added to an air handler or duct run.

Fan Performance

If a packaged forced-air device is involved, manufacturers usually publish values for the "free delivery" CFM associated with each fan speed. No other information is required unless the equipment is designed for use with a duct system. When this is the case, the manufacturer should provide a blower table that documents the relationship between the air-side flow rate (CFM) and the resistance generated by the duct system (IWC).

Unit Sizing

Refer to the manufacturer's performance data and equipment selection recommendations to find a product that provides the required amount of heat. If the unit is equipped with a multiple speed fan, use the performance data that corresponds to the medium speed setting.

Placement

Convective devices should be installed below windows to neutralize the flow of cool air that falls off the inside surface of the glass. If all the capacity cannot easily be installed below the windows, or if there are no windows in the room, the hardware should be installed along the outside wall. If the room is long, the capacity should be distributed (continuously or piece-wise) along the full length of the outside wall. In multi-story structures, a device installed at the bottom of the stairwell will compensate for a cascade of cool air.

Capacity Control

The heating capacity of a device or piping circuit can be controlled by starting and stopping a pump, or by using a two-way valve (throttling or shut-off) or three-way valve (diverting or mixing) to modulate the flow through the circuit. If a device has a multi-speed fan, the capacity can be controlled by changing the fan speed. Or, capacity can be controlled by adjusting the temperature of the water leaving the boiler. In the case of convective devices, capacity might be controlled by adjusting a damper. (More information about capacity control, as it relates to the flow rate and the water temperature, can be found in Section 5-4.)

6-17 Electric Resistance Heat

Electric resistance heat could be supplied by baseboard fixtures, wall convectors, fan-powered unit heaters, or a coil installed in a duct or air handler. Or, electric resistance cables could be installed in a masonry floor or embedded in a ceiling.

Air Pattern Created by Buoyancy Force

If baseboard fixtures or low wall convectors are installed close to the floor on an outside wall, the air motion in the conditioned space will be characterized by a rolling pattern that moves up the wall and around the cross-section of the room. This type of circulation produces comfort at the floor level and limits the floor-to-ceiling temperature difference to about 2°F.

Heating Capacity

The heating capacity of a convector, a fan powered device, or an air-side coil depends on the power consumed by the resistance element, which may be specified as Watts per Foot or Watts, depending on the product. (One Watt translates into 3.413 BTUH of heating capability).

Air-Side Pressure Drop

If a fan-coil device is self-contained, information about the air-side pressure drop across the coil is not required. However, pressure drop information is required when a electric heating coil is added to an air handler or duct run.

Blower Performance

If a self-contained device is involved, manufacturers usually publish values for the "free delivery" CFM associated with each fan speed. No other information is required unless the equipment is designed to work with a duct system. If a fan-coil unit or furnace is designed for use with a duct system, the manufacturer should provide a comprehensive table that documents the relationship between the blower CFM and the external resistance against which the blower will operate.

Unit Sizing

Refer to the manufacturer's performance data and equipment selection recommendations to find a product that provides the required amount of heat. If the unit is equipped with a multiple speed fan, use the performance data that corresponds to the medium speed setting.

Placement

Terminal devices should be installed below windows to neutralize the flow of cool air that falls off the inside surface of the window. If all the capacity cannot easily be installed below the windows, or if there are no windows in the room, the hardware should be installed along the outside wall. If the room is long, the capacity should be distributed, continuously or piece-wise, along the full length of the outside wall. In multi-story structures, a device installed at the bottom of the stairwell will compensate for a cascade of cool air.

Capacity Control

The heating capacity of a resistance heating element typically is controlled by a two-position thermostat (on-off control), but there are electronic devices that can provide proportional control (SCR controllers). Capacity also can be activated in stages if the wiring plan features two or more independent circuits. If a self-contained device has a multi-speed fan, the capacity can be controlled by changing the fan speed. In the case of convective devices, capacity might be controlled by adjusting a damper.

Efficiency

Provided that the device is installed within the conditioned space, the coefficient of performance (COP) of an electric coil is equal to 1.0, which translates into a HSPF of 3.413 BTUH/WATT. If an electric furnace or boiler is installed in an unconditioned space, the COP will be less than 1.0 (but normally more than 0.90) due to the cabinet or jacket loss.

6-18 Packaged Terminal Equipment

Packaged terminal equipment can be used to cool, or heat, a room or zone. Packaged terminal air conditioners (PTAC units) may provide just cooling or they may be equipped with a heating device, which could be an electric coil or a hot water coil. Packaged terminal heat pumps (PTHP units) are usually equipped with an electric heating coil, but some units can be equipped with a hot water coil. Packaged terminal units also can provide a supply of outdoor air capability, which is typically equal to 20 or 25 percent of the blower CFM.

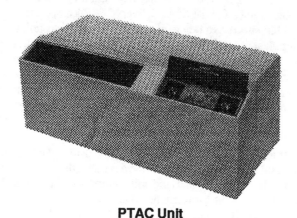

PTAC Unit

Packaged terminal equipment is manufactured to industry grade standards, and it is relatively inexpensive. No duct system is required, but sizable wall penetrations will be required to accommodate the through-the-wall design. (If a supply duct is used, it must not create more than 0.10 inches water column of resistance when the flow is equal to the rated CFM of the equipment.) The newer models have good efficiencies and they provide acceptable control over temperature and humidity; but they have limited air filtration capability, equipment noise is generated at the boundary of the room, and free-delivery of the supply air via an integral grille is relatively crude.

Equipment Selection
Some, manufacturers publish comprehensive performance data. When it is available, this performance data should be used to select and to size the equipment. In any case, when the equipment is subjected to the indoor and outdoor design conditions, the sensible and latent cooling capacity or the heating capacity should be equal to, or slightly exceed, the design load. A discussion of how applied performance data is used to select equipment is provided in **Manual S**.

Controls
The comfort controls are provided with the package. Cooling controls include an integral or remote thermostat and switches that allow the user to select the operating mode, such as on-off, fan-only, cool-low, and cool-high for example. There also may be a knob or lever that allows the occupant to adjust a vent or an outdoor air damper. If the unit has heating capability, additional switches are provided to select one or more heating modes. If safety and operating controls are provided, they usually will be included as standard equipment.

Efficiency Ratings
The efficiency of packaged terminal equipment is characterized either by an EER value or a COP value. These efficiency descriptors are dimensionless numbers that represent the ratio of output energy to input energy. Since they are associated with continuous, full-load operation, they do not indicate the equipment's seasonal efficiency. They also pertain to a specific set of operating conditions (indoor and outdoor dry- and wet-bulb temperatures).

6-19 Window Air Conditioners and Heat Pumps

Window units are classified as appliances, which means that the durability and efficiency requirements are dictated by appliance standards. They are used when low initial cost is significantly more important than precise control over the conditions in the occupied space.

Window units can be used to cool or heat a room or zone. They can provide all the necessary cooling or heating — or they can be used to supplement a central system. They are easy and inexpensive to install, and the newer models offer reasonably good efficiency. Some units have an outdoor air damper and/or a vent damper, and they all have limited air filtration capability. This type of equipment delivers supply air through a grille, which is a relatively crude way to project supply air into the room. In addition, equipment noise is generated at the boundary of the room. Also note that wall sleeves can be used in lieu of a window installation.

When the equipment has heating capability, the heat could be generated by an electric coil, or it could be produced by reversing the refrigeration cycle. During cold weather, electric coils can be used to supplement the reversed cycle output. Capacities range from about 4,000 to 36,000 BTUH — total cooling capacity.

Equipment Selection
Window units do not require much design work. A simple load calculation is all that is needed. Performance data, which is minimal, will provide information about total cooling capacity, heating capacity, and efficiency. In some cases, the manufacturer may provide information about the coil sensible heat ratio (CSHR). This information is desirable because units that have a low CSHR are preferred in a humid climate and units that have a high CSHR are more compatible with a dry climate.

Controls
All the comfort controls are integrated into the package. Cooling controls include a thermostat and switches that allow the user to select the operating mode, such as on-off, fan-only, cool-low, and cool-high for example. There also may be a knob or lever that allows the occupant to adjust a vent or an outdoor air damper. If the unit has heating capability, additional switches are provided to select the heating options. If safety and operating controls are provided, they will be in-

cluded as standard equipment. Usually, there are no alternative control options.

Efficiency Ratings

The efficiency of a window unit is characterized either by an EER value (cooling) or a COP value (heating). These efficiency descriptors are dimensionless numbers that represent the ratio of output energy to input energy. Since they are associated with continuous, full-load operation, they do not indicate the equipment's seasonal efficiency. They also pertain to a specific set of operating conditions (indoor and outdoor dry- and wet-bulb temperatures).

Section 7
Winter Humidification

Humidification during cold weather is desirable because moistened air has a positive effect on comfort and health, because it reduces or eliminates the annoyances caused by static electricity, and because it helps to preserve hygroscopic materials (wood craft, carpets, fabrics, and plaster). However, the amount of moisture in the air must be carefully controlled, or condensation will form on the coldest indoor surfaces — usually on window glass and window frames. Also note that condensation can form on the concealed surface of a structural component — inside a wall or above a ceiling, for example. This type of condensation must be prevented because it can damage the structure, be the source of unpleasant odors, or create an environment that cultivates mold, mildew, and other types of biological growth.

7-1 Desirable Heating-Season Humidity Level

Information about desirable indoor moisture levels, in terms of relative humidity, is provided below. Note that these suggestions pertain only to the physiological benefit for the occupant. In practice, the ambient moisture level also must be compatible with the outdoor design temperature and the construction detail (condensation must not form on any exposed or hidden surface).

- The ASHRAE comfort chart (Figure 1-2) shows the relationship between comfort, relative humidity, and dry-bulb temperature. This chart shows that acceptable heating season humidity levels can range from less than 30 percent to more than 60 percent.

- The ability of the human body, especially the respiratory tract, to resist germs and viruses is optimized when the relative humidity is between 30 and 60 percent. Health and comfort problems associated with the skin and eyes also can be caused by inadequate humidification.

- A humidity level of 35 to 45 percent is required to prevent static buildup on people and on most materials.

7-2 Construction Detail Limits Humidity Level

The upper limit for indoor humidity is established by the construction details and the winter design temperature. In this regard, the designer must evaluate the potential for visible condensation and concealed condensation.

Visible Condensation

Condensation on interior surfaces will not occur if the dew point temperature of the indoor air is lower than the temperature of the coldest inside surface. As indicated by the equation below, this temperature (T_s) depends on the U-value of the structural component, the indoor temperature (T_i), and the outdoor temperature (T_o).

$$T_s = T_I - (0.65 \times U \times (T_I - T_o))$$

For example, the following calculation indicates that if two panes of clear glass have a U-value of about 0.60 (1/4 inch air-space), the temperature at the inside surface will be equal to 42.7°F when the indoor temperature is 70°F and the outdoor design temperature is 0°F.

$$T_s = 70 - (0.65 \times 0.60 \times (70 - 0)) = 42.7\,°F$$

Now, the psychrometric chart can be used to determine the relative humidity associated with a 43°F dewpoint value and a 70°F indoor temperature. As demonstrated by Figure 7-1, the corresponding relative humidity is about 37 percent. This value represents the upper-limit humidity level for this scenario.

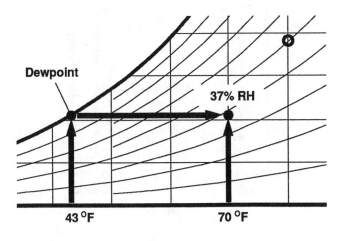

Figure 7-1

Concealed Condensation

Concealed condensation can occur whenever the temperature of a surface that is within a wall, ceiling, or floor is below the dew point temperature of the air that migrates to that surface. If vapor retarders are not used, or if they are improperly installed or damaged after they have been installed, the indoor humidity level that can be maintained without producing concealed condensation can be much lower than the humidity level that would produce visible condensation on the coldest inside surface.

The following equation can be used to evaluate the temperature distribution across a structural sandwich. This equation shows that the temperature at a concealed surface (Tc) depends on the R-value associated with the collection of material between the surface of interest and the outdoors (Rc), the total R-value across the sandwich (Rt), the outdoor design temperature (To), and the indoor temperature (Ti).

$$T_c = T_o + \frac{(T_I - T_o) \times R_c}{R_t}$$

For example, the following calculations compare the condensation potential at the inside surface of a cinder block wall that has insulation installed on the indoor surface with the condensation potential of a similar wall that has the same amount of insulation installed on the outside surface. These calculations are based on the data provided by Figure 7-2, which indicates that the thermal resistance of the cinder block is equal to 2, that the thermal resistance of the path that connects the outdoor surface and the inside surface of the block is equal to 12, and that the R-value of the entire wall is equal to 13. Figure 7-2 also indicates that the indoor temperature is 70°F and that the outdoor temperature is -5°F.

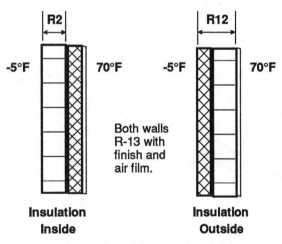

Figure 7-2

Insulation installed at inside surface:
$T_c = -5 + ((70 - (-5)) \times 2)/13 = 6.5\ °F$

At 6.5°F, a low temperature psychrometric chart indicates that dew-point air has 8 Grains of moisture. The standard psychrometric chart indicates an 8 percent relative humidity value at a (70°F / 8 Grain) condition. Therefore, condensation will occur if the indoor humidity exceeds 8 percent.

Insulation installed at outside surface:
$Tc = -5 + ((70 - (-5)) \times 12)/13 = 64.2\ °F$

At 64°F, the psychrometric chart indicates that dew-point air holds 90 Grains of moisture. The chart indicates a 82 percent relative humidity value at a (70°F / 90 Grain) condition. Therefore, condensation will occur if the indoor humidity exceeds 82 percent.

7-3 Moisture Migration

Water vapor rapidly migrates throughout an enclosed space, finding its way into every gap, crack, and cavity that has an air-path connection to the humidified space. This means that, even though humidification may be done locally, the contiguous spaces — as defined by the exterior walls and/or impermeable interior walls — will be humidified. Therefore, the maximum acceptable humidity level in a particular room or zone may be limited by the construction details associated with remote rooms or zones.

Also note that moisture can be mechanically dispersed throughout a home, usually by a forced-air heating system. For example, if return air is drawn from a room that contains a pool or hot tub, the entire home will be humidified. (When the blower is not operating, moisture will migrate through any duct run — supply or return — that connects the humidified room with the other rooms served by the distribution system.)

7-4 Humidification Load

Infiltration loads and mechanical ventilation loads are the most common types of humidification loads associated with residential applications, assuming that the structure is fitted with an effective vapor retarder and ignoring the effect of internal moisture gains. Because these loads are similar, they can be combined into a single load; as indicated by the following equation, this load depends on the total flow of outdoor air (infiltration CFM plus ventilation CFM), the moisture content (Grains) of the outdoor air (at the winter design temperature and 80 percent relative humidity), and the moisture content of the indoor air. (Moisture-content values can be obtained from the psychrometric chart.)

Pounds of water per hour = 0.00064 x Outdoor CFM x ...
... (Indoor Grains - Outdoor Grains)

For example, the following calculations show that 4.77 gallons of water per day will be required to humidify 100 CFM of outdoor air when the outdoor design temperature is equal to 10°F and the indoor design condition is equal to 70°F dry bulb and 30 percent RH. For convenience, Figure 7-3 (next page) summarizes the result of a similar set calculations for a range of outdoor temperatures and indoor humidities.

Outdoor Grains = 7 (10°F, 80% RH)
Indoor Grains = 33 (70°F, 30% RH)
Pounds per Hour = 0.00064 x 100 x (33 - 7) = 1.66
Gallons per Day = (1.66 x 24) / 8.35 = 4.77

Humidifier Capacity Requirement							
Gallons of Water per Day per 100 CFM of Outdoor Air							
Indoor RH	Outdoor Design Temperature - °F						
	-20	-10	0	10	20	30	40
20 %	3.68	3.50	3.31	2.76	1.84	0.55	NA
30 %	5.70	5.52	5.33	4.78	3.86	2.58	0.74
40 %	7.73	7.54	7.36	6.81	5.89	4.60	2.76

Figure 7-3

7-5 Humidification Equipment

Humidification equipment can be classified as adiabatic or isothermal. "Adiabatic" means that no heat is added by the humidification device. This type of equipment is commonly called evaporative equipment. "Isothermal" means that there is no dry-bulb temperature change associated with the humidification process.

Adiabatic humidifying equipment could be an atomizing device, a wetted-element device, or evaporative-pan equipment. Since adiabatic devices do not have a heating element, all the heat that is required to evaporate the water must be extracted from the air being processed by the device. Therefore, this type of equipment decreases the dry-bulb temperature of the humidified air.

As noted previously, there is no dry-bulb temperature change associated with isothermal humidification equipment. In this case, the heat required for the evaporation process is provided by an integral device or by an external heat source. Heated-pan humidifiers and steam humidifiers provide examples of isothermal humidification equipment.

Characteristics of Humidification Equipment				
	Steam	**Heated Pan**	**Atomizing**	**Wetted Media**
Capacity Range (LBS-Water/HR)	1 to 2000 per unit	1 to 20 per unit	6 to 15 per nozzle	4 to 300 per unit
Potential for Biological IAQ Problem	Low — hot steam is sterile, no pan, no drain	Moderate — low capacity, bacteria stays in pan	Very high — water/bacteria sprayed into air	High — bacteria can contact and mix with air
Potential for Chemical IAQ Problem	Low to high, depending on feed-water	Low — dissolved solids accumulate on pan and coil	Dissolved solids will precipitate out of spray	Low to moderate, dissolved solids in the sump
Problems with Water Mist or Drips	None if manifold has jacket	None	Severe to none, depends on design	None
Maintenance Requirement	Annually	By week or month, blowdown is important	By week, month, or year, depends on the design	By week or month, blowdown is important
Heat Source	Self-contained electric heaters or central boiler	Electric, steam or hot water coil; infrared heater	None	None or device to heat water or air heating equipment
Effect on Air Temperature	None	Small rise	Reduction	Reduction
Location	Manifold can be installed in duct or in room	Unit can be installed in duct or in room	Unit installed in duct (residential equipment)	Duct, plenum, or self-contained room unit
Effect on Air Flow in Duct System	Small pressure drop, minimum height requirement	Not a factor when properly installed	Small or negligible pressure drop	Pressure drop; bypass units may reduce supply CFM
Control	On-off or throttling with quickest response	On-off with poor response	On-off with good response, throttling is possible	On-off, response depends on design; slow to good
Installation Considerations	Steam generation, piping, strainers, traps, valves, and condensate return	Energy for heat source, water, and drain piping	Water piping	Water and drain piping, some units may need energy for heat source

Figure 7-4

7-6 Performance Characteristics

On the preceding page, Figure 7-4 compares the performance characteristics of the various types of residential humidification equipment. More information about each performance-related item is provided below. This summary can be used to form a basis for selecting a particular type of humidifying device.

Capacity

Some types of humidifiers are designed to satisfy modest humidification loads and others are designed to handle larger loads. As indicated by Figure 7-5, the capacity of residential humidification equipment ranges from about 1 Lb/Hr (about 3 Gallons/Day) to more than 7 Lb/Hr (about 28 Gallons/Day).

Approximate Capacity Ranges Residential Humidification Equipment	
Steam Humidifiers	
Self-contained unit	1 to 150 Lb/Hr
Single manifold, steam from boiler	1 to 2000 Lb/Hr
Area unit with or without fan	2 to 400 Lb/Hr
Pan Humidifiers	
Water pan in room (1 Sq.Ft.)	0.03 Lb/Hr
Water pan in hot air duct	0.5 to 1.5 Lb/Hr
Plate or disc type - mounted in duct	1 to 3 Lb/Hr
In duct with electric heat element	3 to 4 Lb/Hr/Kw
In duct with hot water coil (160°F)	4 to 6 Lb/Hr
In duct with steam coil (2 PSIG)	10 to 20 Lb/Hr
Self-contained, heated, integral fan	2 to 24 Lb/Hr
Wetted Media	
Wet pad (100 CFM - 75°F / 35% RH)	1 to 2 Lb/Hr
Bypass type - no integral fan	1 to 8 Lb/Hr
Self-contained fan-powered unit	3 to 6 Lb/Hr
Under the duct - rotating media	2 to 10 Lb/Hr
Plenum mounted - integral fan	1 to 10 Lb/Hr
Atomizing Units	
Centrifugal plate or cone	1 to 2 Lb/Hr
Pressure spray nozzle	1 to 10 Lb/Hr
Splashing ring - plenum mounted	1 to 3 Lb/Hr

Figure 7-5

Biological Contamination

Reservoirs, drain pans, dripping or spitting nozzles, condensation on duct materials, or building components are all potential sources of biological contamination. The likelihood of an air quality problem depends on the type of humidification device. In general, sterile feed water, proper sizing and installation, scheduled inspections, and proper maintenance will ensure satisfactory performance.

- Because of the high temperatures associated with steam humidification, there should be no biological contamination — providing the equipment is properly installed and maintained.

- Atomizing humidifiers have the potential to cause health and air quality problems. (They are the most likely to cause surface wetting, they are able to disperse large amounts of bacteria into the air, and problems can be caused by dripping nozzles or stagnate water in drip pans.)

- Evaporative-pan and wetted-media humidifiers can cause the same types of problems as atomizing equipment, but to a lesser extent.

- Heated-pan humidifiers are less troublesome than evaporative devices because they operate at a higher temperature. The vapor produced by these devices is relatively free of biological contaminants, which remain in the pan.

Chemical Contamination

The formation of corrosion, scale, and biological growth can be a problem for any type of hydronic or steam device. Certain types of chemicals are used to retard, prevent, or remove these deposits. However, if these chemicals are present in the humidifier feed water, they can be dispersed into the air. Some of these chemicals are known to produce a health hazard. Therefore, a particular type of humidification device should not be used if there is a possibility that water treatment chemicals will contaminate the humidified air.

Maintenance

Steam and heated-pan humidifiers discharge water vapor into the air. If these devices are properly installed, there should be no scaling or corrosion on duct or building surfaces, but some components of the heated-pan equipment may accumulate deposits.

By comparison, all types of evaporative humidification equipment require scheduled maintenance; especially atomizing equipment, which has a tendency to corrode at the nozzle, and to cause corrosion on duct and structural surfaces. Also, if dissolved minerals are in the feed water, they will precipitate out of the water when atomizer type equipment is used to humidify a space. This causes an accumulation of dust or powder on the surfaces in the room. Even though these deposits do not poise a health hazard, they create a cleaning and maintenance problem.

Heat Source

Evaporative equipment does not have a heat source; it extracts the heat of evaporation from the air that is being humidified. This process causes a drop in the dry-bulb air temperature.

An external heat source supplies the heat-of-evaporation for isothermal (steam or heated pan) equipment. This heat source can be an integral part of the unit, or the heat can be supplied by an external device. For example, a steam humidifier can be self-contained, or steam can be supplied by the boiler used to heat the home (providing that there are no toxic chemicals in the boiler feed water).

Location
Some humidification devices are installed in a duct or plenum. Other types of devices are installed within a room. Duct-mounted units humidify all the spaces served by the air distribution system and any air-coupled spaces not served by the distribution system. Room-mounted units humidify the room where they are located — and intentionally or unintentionally, any adjoining space that is air-coupled with the humidified space. Both types of devices must be mounted so that minimum clearances are maintained and condensation is prevented.

Control
Since the water is already vaporized, steam humidifiers can provide an immediate response to a call for more or less humidification. Other types of equipment have a slower response time because of the time lag associated with the evaporative process. Capacity control also is important. Steam humidifiers and some pneumatic atomizing units provide the opportunity to use modulating controls. The other types of humidifiers are simple on-off devices.

Cost
Equipment costs vary with the size and type of humidification equipment. Installation costs depend on the work associated with installing equipment, installing piping systems, and supplying a heat source. Since the heat required to evaporate a pound of water is the same for every system, the operating cost is determined by the efficiency of the heat source and the cost of the fuel. Steam units tend to have low maintenance costs and atomizing equipment tend to have high maintenance costs.

7-7 Pan Humidifiers

Pan humidifiers are simple devices designed to evaporate water from the surface of a heated reservoir. (If the reservoir is not heated, the device is not very effective.) Heat can be supplied by an infrared lamp or by an electric, hot water, or steam coil. These devices will not cause a significant change in the temperature of the humidified air because the heat of evaporation is supplied by an external source.

Heated-pan humidifiers are suitable for small loads. They can be installed in either a supply air duct or the room. Regular blowdown and weekly or monthly maintenance is required to keep the pan and the heating coils free of biological, chemical, or mineral deposits. When properly installed, there should be no water droplet, wetting, or duct corrosion problems. These humidifiers are on-off devices, and because of dynamics of the evaporation process, they have a sluggish response to a call for humidity or shut-down command.

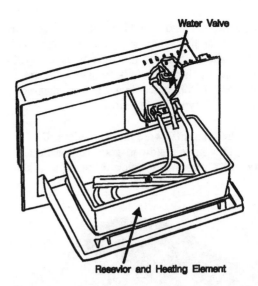

Pan Humidifier

7-8 Wetted Media Humidifiers

Wetted media humidifiers are designed to evaporate water from the surface of a wetted pad. The media can be wetted either by a nozzle or immersion in a sump. (Nozzle units use considerably more water, but they provide continuous blowdown.) Usually, the heat of evaporation is supplied by the air flowing through the media. This psychrometric process causes a drop in the dry-bulb temperature of the humidified air. In some cases, performance may be enhanced by using heated water.

Most wetted media humidifiers are designed for small loads, but some self-contained units have the capacity to handle intermediate loads. This type of equipment is very common, and there are many different types of duct-mounted and self-contained room units from which to choose.

- The air flow through duct-mounted, bypass units is induced by the pressure difference across the supply-air blower. (Bypass units may cause an undesirable reduction in the supply CFM.)

- Some types of duct-mounted devices simply penetrate the duct wall and protrude into the air stream.

- Some types of duct-mounted devices are equipped with an integral fan that circulates air through the media.

- Some duct-mounted designs may dissipate a noticeable amount of static pressure, but some designs do not produce much resistance to air flow.

- The self-contained units designed for installation within a room are normally equipped with a fan, and some of these units may have an integral heat source.

Regular blowdown, and weekly or monthly maintenance is required to clear the sump or reservoir of any biological, chemical or mineral deposits. If wetted media humidifiers are properly installed, there should be no water droplet, wetting, or duct corrosion problems. Mineral fallout (dusting) is not a problem with this type of equipment.

Wetted media humidifiers are on-off devices. Depending on the design, they have a slow-to-good response to a call for humidity and a reasonably quick response to a shut-down command.

Duct-wall, Fan-powered, Wet-media Humidifier

Under-duct, Wet-media Humidifier

Residential Atomizing Humidifier

Residential, By-pass, Wet-media Humidifier

7-9 Atomizing Humidifiers

All types of atomizing humidifiers spray a fine mist of water droplets into the air. This means that the air must supply the heat required to complete the evaporation process. This psychrometric process causes a drop in the dry-bulb temperature of the humidified air.

Residential-type, spinning disk, and diffusing screen units are designed for very small loads. Pneumatically powered, pressurized water, nozzle spray units can be used to satisfy small, medium, and large loads.

Small residential-type units can be installed in a duct, but pneumatic spray nozzles are designed primarily to be installed in the room. There is always the possibility that any type of atomizing equipment could wet nearby surfaces. However, this should not be a problem if the equipment is properly sized, controlled, and installed. (If one of these devices is installed in a duct, the application should be approved by the manufacturer of the atomizing equipment.)

Regular maintenance may be required to clear clogged nozzles of mineral deposits and to clear biological or mineral deposits from the drip pans. Mineral fallout (dusting) is a problem inherently associated with this type of humidification equipment — anything that is in the water, including biological contaminants, will be dispersed into the air.

Small atomizing humidifiers are typically on-off devices, but some manufacturers install modulating controls on models designed to satisfy large loads. In either case, these devices have a relatively quick response to a call for humidity and to a shut-down command.

7-10 Steam Humidifiers

Because of the high temperatures involved, steam humidifiers are biologically the most sterile type of humidification equipment (no algae, bacteria, or odor). However, chemical-related air quality problems have been caused by steam generated from tainted feedwater. For example, if the humidification device receives steam from a central boiler, some of the chemicals used to treat the boiler feedwater can make the steam unacceptable as a source of moisture. Self-contained, electrically-heated units offer a solution to this problem because they can be supplied with feedwater that has acceptable purity. In any case, this type of equipment will not produce a significant change in the dry-bulb temperature of the humidified air because the heat-of-evaporation has been supplied by an external source.

Steam humidification equipment can be installed in a duct or a room. Spitting and water droplets will not be a problem if the unit is equipped with a jacketed manifold and a steam separator. This type of hardware ensures that the steam will be dispersed as a pure vapor. However, even if the steam is introduced as 100 percent vapor, there is always the possibility that condensation could wet nearby surfaces.

If condensation occurs, it will be a source of corrosion or biological contamination, but, condensation and surface wetting should not be a problem if room units and duct-mounted manifolds are properly sized, controlled, and installed. (Units installed in the room are often equipped with a self-contained propeller fan that helps disperse the steam before it condenses.) Note that steam cups and unjacketed steam pipes are not recommended — their performance is characterized by problems associated with droplet fallout, spitting, and wetting.

If the feedwater is of good quality, steam humidification equipment will require a minimum amount of maintenance, perhaps as infrequently as once per year. Dusting and fallout is not a problem with steam equipment.

Steam humidifiers can be controlled by a modulating valve or a two-position valve. In either case, the humidity can be precisely controlled because control valves have a fast response to a call for more moisture and to a shut-down command. (Steam valves must be properly sized to obtain optimum performance.)

7-11 ARI Ratings

Residential, wetted media, and atomizing units are rated according to ARI standards 610 and 620, which specify the conditions for the rating tests. These ratings are for continuous operation. Capacities are listed in *Gallons-per-Day* for a specific set of operating conditions, which include air velocity, dry-bulb and wet-bulb temperatures, entering water temperature, and water flow rate. Some examples of these rating conditions are provided below.

Standard 610 — Central Units
- Capacity listed in gallons per day
- Air velocity in duct = 800 feet per minute
- Supply air temperature = 120°F
- Return air temperature, duct/plenum = 75°F
- Relative humidity, return duct/plenum = 30 percent
- Entering water temperature = 50 to 60°F

Standard 620 — Self-Contained Units
- Capacity listed in gallons per day
- Entering air temperature = 75 °F
- Entering humidity = 30 percent
- Entering water temperature = 60°F (no heat)
- Entering water temperature = 120°F (heated)

Note that when the actual operating conditions are different from the rating conditions (a typical situation), the installed capacity will not be the same as the rated capacity. Unfortunately, manufacturers do not usually publish comprehensive application data for small units. For example, Figure 7-6 shows how capacity is affected by dry-bulb temperature.

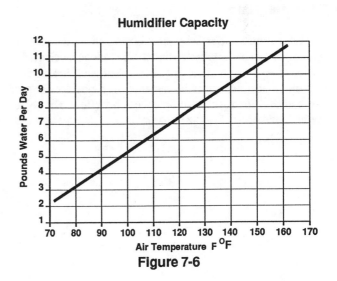

Figure 7-6

7-12 Application Data

Application data differ from rating data because it provides capacity information for a variety of operating conditions.

This information is not normally published for small-capacity devices, but it should be used to select and size humidification equipment when it is available.

7-13 Controls

The output capacity of a humidification device can be controlled by a humidistat that senses the amount of water vapor contained in room air, or the sensor can be located in a duct. The humidifier manufacturer may provide the humidistat as standard equipment or as an option, or it may be obtained from another source. When a duct-mounted humidifier is used, the humidistat should be interlocked with the blower. This interlock synchronizes operation with the fan and prevents wetting when there is no air flow. In addition, a high-limit humidistat should be installed in a duct or in the room to protect against condensation and wetting. In some designs, a temperature switch is used to lock out the humidifier during a morning warmup cycle. If the unit has a sump or reservoir, a high water cut-off valve and a bleed-off valve/drain may be included as standard equipment or as an option.

Steam humidification equipment manufacturers offer various types of control valves and valve operators, which may be modulating or on-off devices. All control valves and operators must be sized correctly.

The wetted-media and atomizing-nozzle devices that are compatible with residential applications normally include capacity control hardware. In some cases, a regulator may be required to control the entering water pressure.

7-14 Installation

Any type of humidifier, whether it is mounted in a duct or in the room, must be installed so that there is adequate clearance between the device and the neighboring surfaces. In addition, duct-mounted units must be installed at a position free of upstream and downstream turbulence or backflow. The objective is to make sure that the moisture is completely vaporized and mixed with the ambient air before it contacts any surface. Failure to provide proper clearance can cause wetting and condensation problems. These off sets are documented in the manufacturer's literature and installation manuals.

Depending on the type of humidifier, it may be necessary supply the device with water or steam. Other installation could involve drain pans, drain lines, condensate return li pipe insulation, strainers, steam traps, water treatment eq ment, electrical service, and controls. Information about requirements is usually provided in the manufacturer's t nical literature and installation manuals.

7-15 Summary of the Design Procedure

The same design procedure can be used to select and size type of humidification equipment. The basic steps of this procedure are summarized below.

1) Establish the indoor humidity and temperature levels that are appropriate for the application (Section 7-1).

2) Establish the maximum humidity level the structure can tolerate (Section 7-2).

3) Finalize the design value for the humidity level. Base this value on the required humidity level (Step 1), providing that it is compatible with the humidity level that can be tolerated by the structure (Step 2). If there is a conflict between the required humidity level and the humidity level that can be tolerated by the structure, alter the thermal performance of the structure or design for a lower humidity level.

4) Use the design value for the humidity level to calculate the net load on the humidification equipment (Section 7-4).

5) Use manufacturer's performance data to size the humidification equipment.

6) Specify controls and accessories.

Section 8
Filters

Indoor air quality depends on the amount of particles that are suspended in the air and on the type of odors, fumes, and gases that are mixed with the air. Examples of particle-based pollution include dust, pollen, spores, tobacco smoke, cooking smoke, animal dander, bacteria, viruses, skin flakes, and carpet fibers. Some of this contamination can be seen as it drifts through a shaft of sunlight, but visible particles only represent one to two percent of air borne contamination. The remaining portion, while not visible to the human eye, becomes evident when it soils or discolors structural surfaces, furnishings, and drapes. The invisible spectrum of particles and pollutants can also be detected by smell or by the undesirable physiological reactions that the human body has to an offensive agent.

8-1 Controlling Contaminates

Indoor air pollution can be controlled by removing the source of contamination, by mixing clean air with contaminated air (dilution), and by filtering the air. Given these options, source-removal is the preferred strategy, but this approach can be used only when the offending substance is generated or emitted at a specific location subject to the authority of the homeowner. This means that the offending agent must be diluted or mechanically removed when the source of contamination is significantly distributed, ubiquitous, or generated off-site. However, the dilution option is only viable when there is an inexpensive source of unpolluted air, which often is not the case for many urban and rural home sites. (The outdoor air could have a high concentration of particulate material, or it could be contaminated by the gases and fumes associated with urban environments, certain types of industrial sites, and feed-lot operations). Also note that filtration (mechanical removal of particulate) provides only a partial solution to air quality problems because filters are not effective on odors and gases. This means that a systematic approach is required to provide acceptable air quality, as outlined below.

- Remove the source whenever possible.

- Use exhaust equipment to control the source.

- Use outdoor air to satisfy a combustion air requirement.

- Use outdoor air to satisfy a minimum ventilation requirement.

- Use a filter to control particle populations.

- Make sure that the heating and cooling equipment has the capacity to process the flow of outdoor air.

- Make sure that the blower has the power to overcome the resistance of the filter.

- Special air-cleaning or air-washing equipment may be required to remove odors and small particles (non-standard application).

8-2 Filtration Mechanics

Filters can be designed to take advantage of two types of cleaning techniques: a straining action and a capture process. If the design emphasizes straining, the openings in the media must be smaller than the objectionable particles. Unfortunately, this approach can create clogging and pressure-drop problems when used to remove small particles. However, these air flow problems can be avoided by using a filter design that emphasizes a capture process. (Capture occurs when particles collide with the fibers of a media or stick to the plates of an electronic air cleaner).

8-3 Measuring Particle Size

Researchers use "microns" to measure the size of airborne particles. This the unit of measurement is relatively small (one micron equals 0.0000393 inches), but under ideal conditions the human eye can see a 10 micron particle. (The dots above the "i's" in this paragraph are about 400 microns in size.)

8-4 Particle Distributions

Figure 8-1 shows that the best set of eyes will not detect 99.9 percent of the population of airborne particles, because only one-tenth of one percent — by count — of the levitating minutia are larger than 1 micron in diameter. However, this one-tenth of a percent is significant because it represents 76 percent of the suspended material by weight; which means that a few dirigibles are mixed in with the microscopic flotilla.

Particle Distributions by Count and Weight		
Size in Microns	By Count	By Weight
0.01 — 0.10	70 %	0.10 %
0.10 — 1.0	29.9 %	20 %
1.0 — 10	0.10 %	76 %
Above 10	Negligible	About 4 %

Figure 8-1

8-5 Particle Size and Filter Effectiveness

The effectiveness of air cleaning devices vary with particle size. For example, the efficiency of dust, lint, and static-charge filters drops rapidly when particle sizes are less than 100 microns, and they are essentiality useless when the particle diameter is less than 10 microns, which means that these types of filters have a negligible effect on air quality.

Figure 8-2 summarizes the relationship between particle size and filter performance. This diagram shows that an ASHRAE dust-spot efficiency of about 10 percent will provide dirt-loading protection for air-side devices (coils and fan blades) and some relief for individuals who have allergies. The figure also indicates that an ASHRAE dust-spot efficiency of 35 percent or greater is required to provide a significant amount of respriable (10 microns or less) particle removal.

8-6 Settling Rate

Particle settling rates depend on particle size. Small particles — about 1 micron or less — can remain suspended indefinitely, but particles that have a diameter in excess of 5 microns tend to settle out of the air in 20 minutes or less. This means that some of the visible particles may never reach the air cleaner, especially during part-load conditions when the blower is operating intermittently.

8-7 Measuring Filter Efficiency

A number of tests are used to measure filter effectiveness, and each test serves a different purpose. As might be expected, some tests produce data that is more useful, for comparison purposes, than others. The following comments apply to the

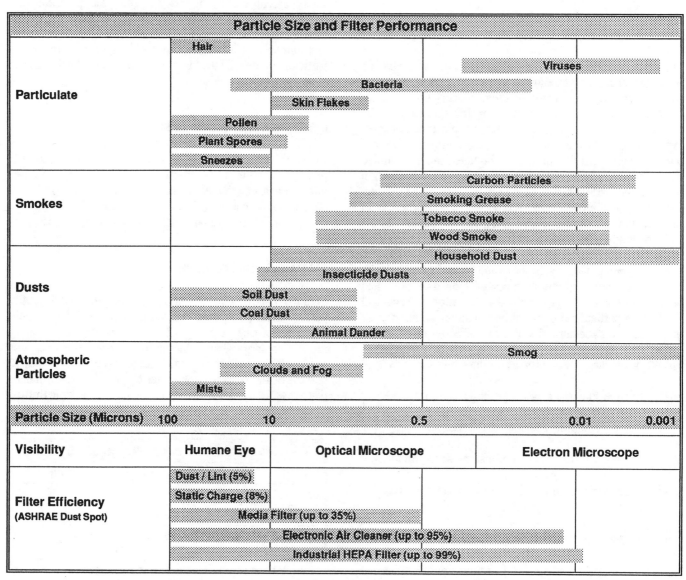

Figure 8-2

tests that can be used to evaluate the filters and air cleaning devices associated with residential applications.

Weight Arrestance Test

Synthetic weight arrestance tests measure the total weight of particles removed by the filter over a fixed period of time. This means that a high arrestance rating can be awarded to a filter that effectively captures a relatively small number of large, heavy particles. (The one-tenth of a percent of the particulate population that is larger than 1 micron accounts for about 80 percent of the captured-particle weight.) Therefore, this index does not provide information about the filter's ability to remove the tiny particles that cause indoor air quality problems.

Atmospheric Dust-Spot Test

The atmospheric dust-spot test measures the filter's ability to remove a wide spectrum of particles. This means that a dust-spot efficiency rating can be used to evaluate filter effectiveness, as it pertains to improving the indoor air quality. (The dust-spot test does not pertain to air quality problems caused by odors and gases.)

Figure 8-2 indicates that a dust-spot efficiency rating of about 10 percent is required to capture the larger particles that cause allergic reactions (pollen, spores, and hair). It also indicates that a dust-spot efficiency rating of at least 35 percent is required to eliminate a wide spectrum of troublesome particles, and that an efficiency rating of at least 90 percent is required for comprehensive particle removal. A high dust-spot efficiency rating (60 percent or more) also indicates that the filter will be able to capture some of the particles that stain and discolor ceilings, walls, and fabrics.

Dust-Holding Capacity Test

The dust holding capacity test measures the amount of dust (by weight) that can accumulate on the filter without causing the air-side pressure drop to exceed a limiting value that is compatible with the intended use. The information is useful for comparing the maintenance requirements of various types of filters.

Fractional Efficiency Test

ASHRAE is developing a new standard for measuring filter performance. When this standard is finalized and adopted, the dust-spot efficiency test could be replaced by a fractional efficiency test. This test will be designed to measure the filter's ability to remove respirable particles in the 0.30 to 10 micron range.

8-8 Air-Side Pressure Drop

The pressure drop across a clean air filter depends on the porosity of the element and the velocity of the flow. As might be expected, an increase in media density or an increase in velocity will translate into a larger pressure drop. This behavior is summarized by Figure 8-3, which compares the perfor-

mance of various types of filters over a range of velocities. Notice that as the dust-spot efficiency of the filter increases, there is a corresponding increase in the pressure drop across the filter. This characteristic means that high efficiency media filters are not compatible with air-handling equipment that has a limited amount of blower power, which is often the case in residential applications. The solution to this problem also is indicated by Figure 8-3, which shows that an electronic air cleaner has a very effective efficiency rating and a relatively low pressure drop.

Air-Side Pressure Drop Exhibit				
CFM	FPM	Flat Viscous Media (10%)	Pleated Dry Media (30%)	Electronic (65%)
800	267	0.02	0.05	0.03
1,000	333	0.03	0.08	0.04
1,200	400	0.04	0.12	0.06
1,400	467	0.06	0.16	0.08
1,600	533	0.08	0.20	0.10
1,800	600	0.10	0.26	0.13
2,000	667	0.12	0.32	0.16

1) Approximately 3 square feet of face area (FPM = CFM/3.0)
2) Randomly selected samples. Refer to manufacturer's data for performance information about specific products.

Figure 8-3

Most furnaces, heat pumps, and air conditioners are tested and rated with a clean filter installed. When this is the case, the manufacturer's blower data does not have to be corrected for the pressure drop across the filter. On the other hand, a correction will be required if the blower is tested with no filter in place, or if another type of air cleaning device is substituted for the one that is supplied with the air handling equipment. (Refer to **Manual D** for more information on available pressure calculations.)

8-9 Dirt Loading

With use, filters either become surface-loaded (particles accumulate on the upstream surface), or depth-loaded (particles accumulate within the filter media). In either case, the accumulation of dust and dirt affects the air-side pressure drop and filtering efficiency. In this respect, electronic air cleaners and media filters have different reactions to an accumulation of dust and dirt.

• Dirt loading does not have a significant effect on the pressure drop across an electronic filter, but the efficiency will slowly decrease as the thickness of the coating increases.

• The pressure drop across a media filter increases as it loads with dirt; if the filter is not changed, the resulting reduction in airflow will degrade comfort and strain mechanical equipment. Dirt loading also may cause a small efficiency improvement (because porosity is decreased), but this effect is insignificant and irrelevant.

8-10 Types of Filters

Filters can be classified according to the method used to remove airborne particles. The design could utilize media coated with a viscous material, or a dry media, or it could feature a charged-particle capture process. Filters also are identified by the contour of the media material, which can be characterized as flat, pleated, or bagged.

Viscous Impingement Filters

Viscous impingement filters use sticky coatings to increase the capture capability of a coarse matrix of flat media material — typically glass fibers, expanded metal mats, animal hair, or nylon thread. These filters can sift large particles from the air, but most of their effectiveness is attributed to impaction captures by the tacky material. (Impaction captures occur when small, non-strainable particles cannot change direction fast enough to follow the circuitous flow-path associated with the trip through the media.)

Viscous impingement filters are characterized by low dust-spot efficiencies and relatively small air-side pressure drops. This means that this class of filters will not have a significant effect on the type of particles that affect indoor air quality. The primary function of this product is to keep large particles of dust and lint from coating the surfaces of the air-side equipment. As indicated by Figure 8-4, the dust-spot efficiency of viscous impingement filters depends on the thickness of the panel (weight arrestance values are included so the reader can compare the two efficiency indexes).

Performance of Viscous Impingement Filters		
Thickness	**Dust-Spot Efficiency**	**Weight Arrestance**
1	5 to 10	20 to 50 percent
1 to 1-3/4	5 to 10 percent	50 to 75 percent
1-3/4 to 2-1/2	5 to 20 percent	60 to 80 percent
2-1/2 to 4	10 to 25 percent	70 to 85 percent

Figure 8-4

Dry Media Filters

Dry media filters clean the air by straining out the particles that are too large to pass between the openings in the media and by impaction. With this design, a large surface of microscopically porous media is required to maximize useful filter life, and to minimize the pressure drop across the filter. This desire for a large surface area can be reconciled with the need for a reasonable frame size by arranging the media in a pleated pattern within the frame. This means that the depth of the filter will depend on the amount of surface area installed within a given frame size. If the filter is relatively thin, for example two inches deep, the surface area will be relatively small, and the efficiency, as defined by media porosity, will be limited by a maximum pressure drop requirement. So, for a particular type of media, the depth of the filter will depend on the desired efficiency requirement and a pressure drop limitation, which is why some high efficiency filters are six or more inches deep.

Electronic Air Cleaners

As airborne particles move through an electronic air cleaner, they lose electrons in the ionizing section and acquire a positive charge. As they proceed through the device, these positively charged particles are attracted to a set of negatively charged plates. Upon contact with the plates, the particles lose their charge and are held in place by the sticky coating that covers the plate.

Electronic air cleaners are very compatible with residential air handling equipment because these devices combine a relatively high dust-spot efficiency with a low air-side pressure drop. As might be expected, the efficiency decreases as the plates collect dirt and dust, but this process does not have a significant effect on the pressure drop. (Dust loading is not a serious problem because efficiently levels can be maintained by periodic maintenance.)

Charged-media Filters

Charged media filters use the principle that governs the operation of the electronic air cleaner, except that an electrostatic charge is applied only to the media. As might be expected, the attraction between the activated media and the un-ionized particles is weaker than the positive-negative attraction associated with ionized-particle—negative-plate devices. Therefore, these devices are not as effective as the electronic air cleaners.

Plastic, Static-charge Air Cleaners

Some filters are made of a plastic material that generates a static charge when air is forced through the media with adequate velocity. This rubbing action produces a mild electrostatic field that enables the media to capture some of the smaller particles not easily captured by straining or impaction. However, this effect is not nearly as powerful as the process associated with an electronic air cleaner (dust-spot efficiencies are less than 10 percent), and it becomes weaker as the humidity of the entering air increases.

Plastic-foam Filters

The open-cell media found in plastic-foam filters is not usually coated with a viscous substance. This means that large particles must be removed by a straining action, and that small particles must be captured by the impaction process. Since the effectiveness of this capture technique depends on the porosity and thickness of the media, the dust-spot efficiency of these devices is characteristically low.

8-11 Filter Maintenance

Dirt-loaded filters must be either changed (media type) or washed (electronic units and some metal media devices) periodically. In some cases the density of the media filter may vary across the depth of the filter, or it may have a facing on one side. This is done to encourage depth loading, so when these types of filters are changed, they must be installed so that air flows in the direction recommended by the manufacturer.

8-12 Controlling Odors and Gases

Media filters (viscous coated or high efficiency), electronic filters, and electrostatic filters do not have an effect on air quality problems caused by odors and gasses. This means that an alternative method will be required to control this type of contamination. In this regard, molecular contaminates could be removed by an activated carbon filter, a catalytic converter, an absorption scrubber, a bed of adsorbent material, or an incinerator. Unfortunately, this technology was not designed for residential applications. (Consider the air-side pressure drop, installation and maintenance requirements, and installation and operating costs.) However, there are a few products that deserve consideration.

Activated Charcoal Filters
Activated charcoal filters can be purchased in panel configurations that are one or two inches thick. Although these products are dimensionally similar to conventional filters, they have a much larger pressure drop when they are new, and as indicated by Figure 8-5, this pressure drop increases dramatically as the filter loads with contaminates. Obviously, pressure drops in this range are not compatible with the blowers that are supplied with residential equipment.

Charcoal Filter Pressure Drop (IWC)*			
	300 FPM	500 FPM	Final
Low	0.15	0.30	1.0
High	0.35	0.75	1.5 to 2.0

* Exhibit values are not generic. Refer to manufacturer's data for information about a specific product.

Figure 8-5

Self-contained Air Scrubbers
Small self-contained air cleaners can be purchased from a number of suppliers. These appliances could feature a charcoal filter, a sorbent material, an oxidizer, or some combination of mechanical and chemical devices. To analyze performance, the contractor should gather information from as many suppliers as possible before selecting a device.

Section 9
Controls

Operating controls are part of a larger system that includes the structure, the fuel conversion equipment, and the distribution system. If properly selected and adjusted, operating controls allow the comfort conditioning system to operate at its full potential, but they cannot be expected to compensate for an inappropriate design concept, improper equipment sizing, or questionable installation practices. Homeowners are generally familiar with operating controls because they use them on a day-to-day basis.

Although safety controls have no effect on comfort, they are just as important, or perhaps more important, than the operating controls because they protect the equipment and the structure. However, homeowners may be totally unaware of these devices because they normally are adjusted by the installer or the service technician. Therefore, contractors should make sure that the homeowner understands which safety and supervisory controls are required, why they are required, and how they affect the installation cost.

9-1 Thermostats

Residential thermostats are electrical switches that activate or deactivate the fuel conversion equipment in response to local changes in indoor air temperature. If the home is not zoned, the thermostat should be installed in the area that is most compatible with the time-of-day and time-of-year load patterns associated with the various rooms. If the home is zoned, one thermostat will be required for each zone and the operation of these thermostats should be monitored by a central control panel. (Refer to Section 3 for more information on zoning).

Thermostats should be installed about five feet above the floor, in a location that will not be affected by solar heat flows through windows, convective or radiant transfers from fireplaces, or convective exchanges from supply outlets or appliances. A thermostat also should be located in a position that is not influenced by drafts caused by door traffic.

Separate thermostats may be used for heating and cooling, or one combination thermostat may be used to regulate the indoor temperature during any season. If a combination thermostat is installed, it usually will allow the operator to select the heating function, the cooling function, or the auto-change-over feature.

Most thermostats have an auto-manual fan switch. This feature is desirable for single-zone systems because continuous fan operation improves room-to-room temperature control (refer to Section 3 for more information on this subject).

Thermostats should be evaluated on their ability to control the local air temperature. The best designs can limit local temperature variations to one or two degrees. (Close control over the local air temperature may not solve a zoning problem, but it might help reduce the size of the temperature excursions in the isolated areas.)

Some thermostats have programmable features that allow setup and setback schedules. In this regard, some models are easier to program than others, and some allow more scheduling options. Therefore, when selecting a programmable thermostat, it is necessary to consider the number of setpoint-changes-per-day and the number of independent daily-schedules-per-week.

Single-stage thermostats are used when the fuel conversion equipment is desigend to cycle between full capacity and zero capacity. Multiple-stage (usually two-stage) thermostats are used when the fuel conversion equipment has the capability to provide full capacity, and one or more steps of intermediate capacity.

9-2 Gas Furnace Control

When a gas furnace thermostat senses a need for heat, it initiates a sequence of control-cycle events that are product specific. The devices that require supervision could include the gas valve, an ignition device, the air-handling blower, a vent fan, a vent damper, a combustion air blower, a high-limit switch, a low-limit switch, a time-out device, and interlocks. (On older designs, the ignition systems normally utilize a standing pilot and a thermocouple to control the gas-train.)

Gas Valves
Traditionally, single-stage gas valves have been supplied with residential equipment, but multi-stage valves are featured in some product lines. (Staging is desirable because comfort is enhanced when the heating capacity of the furnace is compatible with the instantaneous heating load.)

High-Limit Control
The high-limit control prevents furnace overheating, which might be caused by a faulty thermostat, dirty filters, burner over-firing, or throttled supply air outlets. This control is a safety device that does not function except under abnormal conditions.

Air-handling Control
A fan control is used to energize the blower motor when warm air is available. In the auto-mode this device may operate independently of the fuel train controls, or it may be se-

quenced with other control actions. It also is possible to activate the blower by using a manual switch that is included with the thermostat. If the equipment package includes a speed control option, the control package will stage or modulate the speed of the blower motor.

9-3 Oil Furnace Control

When an oil furnace thermostat senses a need for heat it activates a control sequence that operates the oil burner motor (which supplies oil and air in the proper mixture) and energizes the ignition system. This process is monitored by safety controls, which will stop the oil burner motor and the ignition device if the burner fails to fire. (The limit and fan controls supplied with an oil furnace perform the same functions as the gas furnace controls.)

9-4 Controlling Electric Resistance Heat

When the thermostat senses a need for heat, it activates a bank of electric resistance heating elements, which may be located in baseboard fixtures or in a central air handler. If an air handler is involved, the blower motor also will be energized. Capacity control can be provided by energizing banks of resistance elements in stages (a desirable strategy). If an electric furnace is featured, the limit controls and fan controls perform the same function as in a fuel-fired unit.

9-5 Hot Water Heating System Controls

The schemes that are used to control the components of a hot water heating system are documented in hydronic system design manuals. In this case the control system must monitor the performance of the ignition system, the fuel-train equipment, the temperature of the boiler water, one or more water pumps, and possibly one or more water valves.

9-6 Cooling Equipment Controls

When the cooling thermostat detects a load, it activates the compressor motor, the condenser fan motor (air cooled equipment), a pump (water cooled equipment), and the indoor blower motor. When the equipment is operating, the refrigerant-side performance should be monitored by high- and low-pressure controls (to protect the compressor), and the operation of the compressor motor should be subject to the approval of a safety device that monitors winding temperature or the motor current. When air-cooled equipment is idle, an undesirable migration of refrigerant can be prevented by activating a crankcase heater.

Low-Pressure Control
The low-pressure control protects against operational problems or compressor damage that could result from a loss of

refrigerant, or a freeze-up caused by insufficient air flow through the evaporator coil.

High-Pressure Control
The high-pressure control is designed to prevent compressor damage if a high-discharge pressure condition is caused by insufficient heat transfer at the condenser. (With air-cooled equipment, the air flow could be blocked by an accumulation of debris on the coil, or a fan malfunction. With water-cooled equipment, the problem could be due to fouling or blockage in a pipe, fouled condenser tubes, or a pump failure.)

Anti-short-cycling Device
Anti-cycling devices protect the refrigeration-cycle equipment if a system malfunction produces an operating condition that is characterized by rapid series of restart attempts.

Low Ambient Lockout
A low ambient control (air-source equipment) prevents operation if the outdoor temperature falls below a value that could damage the refrigeration-cycle equipment. (This is normally not a problem in residential applications.)

Hard-start Kit
A hard-start kit provides more starting power at the compressor motor, but this device may be a solution to a problem that will not exist if the equipment is adequately controlled and functioning properly. (The crankcase heaters should keep bearings warm during cold weather off-cycles, and the anti-short cycle relay should allow the low- and high-side pressures to equalize before restart. Pistons must not be binding in the cylinders, bearings and motor windings must be in good condition, valves must be operating properly, and there must be no blockage in the refrigerant lines.)

Blower Control
In the auto-mode, the blower will be activated along with the compressor. If continuous blower operation is desired, the blower may be energized by a manual switch.

9-7 Heat Pump Controls

Most of the controls supplied with heat pump equipment are similar to the controls used for other types of refrigeration-cycle equipment. However, there are some controls that are peculiar to heat pump applications. These devices are associated with the defrost cycle (air-source equipment), freeze protection (water-source equipment), and operation of the auxiliary heating device (air- or water-source equipment).

Defrost Cycle
Unless the ambient air is extremely dry, the outdoor coil (air-source equipment) will accumulate a coating of ice when the heat pump is operating in the heating mode. Since this cladding degrades performance, it must be removed. Usually, this is accomplished by reversing the operating cycle of the heat pump (heat extracted from the indoor air and rejected through

the outdoor coil) and activating the auxiliary heat. This mode of operation is initiated and terminated by the action of the defrost control, which is normally a time-temperature device or an upon-demand device. (As explained in Section 3 of **Manual H**, demand defrost is significantly more efficient than time-temperature defrost because there is a direct relationship between the initiation of a defrost cycle and an icing problem.)

Freeze Protection
If water-source equipment is featured, a low water temperature cut-out (freeze stat) can be used to protect the refrigeration-cycle equipment if the water temperature is too low.

Supplemental Heat
During normal operation, the supplemental heat (electric coils or fossil fuel furnace) is activated by the second stage of the indoor thermostat. In addition, an outdoor thermostat should be used to lock-out the supplemental heat if the outdoor temperature is above the thermal or economic balance point. Also note that the load on the utility power-grid can be minimized, and the supply air temperature can be optimized, if electrical resistance heat is controlled so that banks of coils are activated in stages as the outdoor temperature decreases. (In **Manual S**, Appendix 7 explains why staged, resistance heat improves occupant comfort during the heating season.)

Emergency Heat
If resistance heating coils are included with the heat pump package, the emergency heating capacity is equal to the difference between the total installed resistance-coil capacity (auxiliary heat) and the total second-stage capacity (supplemental heat). This increment of capacity should not be available during normal operation, but if the refrigeration-cycle equipment fails, it can be activated by a manual emergency heat switch.

Section 10
Air System Design

Comprehensive design work is a prerequisite to a successful installation. This work begins with the conceptual task of selecting an appropriate system and progresses through a series of procedural exercises that pertain to selecting, sizing, and arranging the system components.

10-1 Design Procedure Flow Chart

Figure 10-1 provides a flow chart of the residential design procedure. This diagram shows that the creative process involves six interrelated steps, which are discussed below.

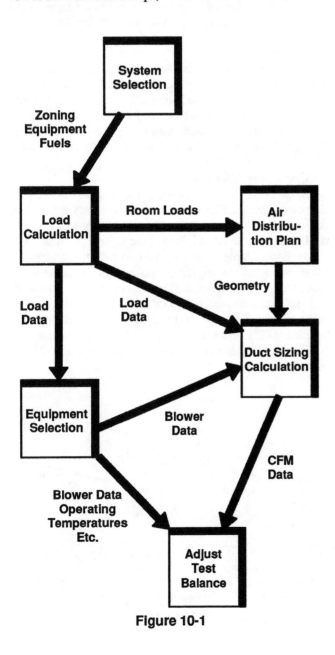

Figure 10-1

10-2 System Selection

One of the first conceptual issues that the designer must address pertains to zoning. Multiple levels, complicated floor plans, and elegant architectural treatments should translate into a design that features multiple, single-zone systems or central, multi-zone equipment. (Section 3 of this manual discusses the construction features that make it difficult for a single-zone system to control room-to-room temperature differentials.)

Other conceptual issues are associated with function, fuel availability, space requirements, noise, and esthetics. Functionally, the system could be a simple heating-only system; or at the other extreme, the conditioning equipment could provide heating and humidification, cooling and dehumidification, ventilation (outdoor air), and a generous amount of filtration efficiency. The design concept also must be compatible with the on-site fuel options, the space available for equipment, and the space available for duct runs and air distribution hardware. In addition, the system must not be a source of objectionable noise — for the occupants or neighbors — and the terminal devices must be compatible with the architecture and interior decor of the home. The difficulties associated with obtaining combustion air and routing vents also should be considered.

Another type of system selection problem is created when the heating and cooling capacity of the equipment is not compatible with the relative size of the heating and cooling loads. A classical example of this scenario is provided by a home located in a sub-tropical climate. In this environment, the heating load is relatively small compared to the cooling load, which means that a furnace sized to the heating load may not have enough blower capacity for cooling. (In this situation, a heat pump is favored because this type of equipment offers a better balance between heating and cooling capacity and the relative size of the seasonal loads.)

The system selection rubric also is subject to economic considerations. Typically, homeowners desire a design that minimizes installed cost, minimizes operating and maintenance cost, and maximizes comfort. Unfortunately, these goals are not compatible, because at the margin an improvement in operating efficiency, or level-of-comfort, translates into an increase in installed cost. Therefore, in order to produce a design compatible with the client's budget, the designer and the homeowner must work through a series of compromises regarding system efficiency and delivered comfort. (During this process, the designer should check to see if a utility incentive program will affect the installed cost, financing cost, or operating costs.)

Occasionally, a home may have one or more features that require a design outside the purview of the traditional design procedures — an indoor pool, hot-tub, solarium, or atrium, for example. When these opportunities present themselves, the designer should refer to the procedures that apply to commercial applications or to the procedures used to select specialized equipment (swimming pool heat pumps, for example).

10-3 Load Calculations

The flow chart of the residential design procedure shows that all equipment selection and sizing decisions, and all air-side design decisions, are based on the information generated by the **Manual J** load calculation. Therefore, the load calculation should be the product of an enthusiastic effort to obtain the most accurate answer possible. Some suggestions for achieving this goal are provided below.

Outdoor Design Conditions
The outdoor design conditions tabulated in **Manual J** (Table 1) do not represent the worst weather conditions ever experienced in the listed city; but they do represent extremes that, on the average, will not be exceeded for more than a few dozen hours per season. This means that a design based on the recommended design temperatures will be optimized for the thousands of hours associated with moderate weather patterns (see Figure 10-2). However, this concept must be explained to homeowners, because as a group they do not understand that installation and operating cost will increase, and that comfort will be compromised if the equipment-sizing calculations are based on record-setting weather conditions.

Indoor Design Conditions
Manual J recommends a 70°F indoor design temperature for heating and 75°F for cooling. The indoor humidity also affects comfort. In the winter, the maximum relative humidity value will depend on the potential for unwanted condensation on cool surfaces (typically, the inside surface of a window). In the summer, the design calculations should be based on a relative humidity of 55 percent or less (preferably 50 percent or less). These recommendations are compatible with the ASHRAE comfort chart (see Figure 1-3).

Design Conditions — Authority
The outdoor or indoor design condition may be dictated by a code or a utility regulation. In this case, the statutory requirements supersede the **Manual J** recommendations or a homeowner's opinion. If there is no codified mandate, the contractor and the homeowner should mutually agree on the design conditions. During these negotiations, the contractor should make a determined effort to educate the homeowner about the disadvantages of oversized equipment.

Infiltration Estimates
With reasonable effort, the winter design infiltration rate associated with an average-size home (1,500 to 2,500 Sq.Ft. of floor area) can be held to less than 0.30 air changes per hour (ACH). Full credit should be taken for this type of construction when it can be justified by an inspection of the construction detail, a blower door test, or the builder's track record.

Air Change per Hour Values
ACH values are meaningless unless they are referenced to the size of the home. This correlation is necessary because an ACH value is generated by an equation that processes information about the leakage CFM and the volume of the enclosed space. For example, consider two homes that are identical, except for a 2:1 scaling factor applied to the wall area. As indicated by Figure 10-3, the larger home has twice as much leakage area, which translates into a commensurate increase in the leakage CFM, but the increase in volume is more than doubled, which means that the larger home will have a smaller ACH value. (The **Manual J** infiltration estimate accounts for this effect.)

Hourly Temperature Summary — Atlanta, GA (Design Conditions ... 22°F Heating; 92°F Cooling)			
Temperature	January	July	Annual
Above 94	0	7	21
90 - 94	0	40	135
85 - 89	0	91	367
80 - 84	0	143	612
75 - 79	0	171	839
70 - 74	5	234	1201
65 - 69	18	55	986
60 - 64	47	6	845
55 - 59	69	0	773
50 - 54	89	0	709
45 - 49	106	0	665
40 - 44	117	0	608
35 - 39	123	0	471
30 - 34	91	0	303
25 - 29	46	0	134
Below 25	30	0	85

Figure 10-2

Floor 40' x 30' = 1200
Wall 140' x 8 ' = 1120
Infiltration = 100 CFM
Volume = 1200 x 8 = 9600
ACH = (100 x 60) / 9600 = 0.62

Floor 80' x 60' = 4800
Wall 280' x 8 ' = 2240
Infiltration = 200 CFM
Volume = 4800 x 8 = 38400
ACH = (200 x 60) / 38400 = 0.31

Figure 10-3

Blower Door Tests
A blower door test does not simulate the pressure differentials and flow patterns caused by wind, indoor-outdoor temperature differences, or pressure differentials produced by mechanical exhaust systems and fuel burning equipment. (Refer to **Manual D**, Section 13 for a comprehensive discussion of the pressure drivers that operate on a home.) This means that a measured blower-door flow rate value should never be interpreted as an estimate of the infiltration CFM. However, researchers have published equations and calculation procedures that can be used to convert blower-door data into an estimated infiltration rate. Unfortunately, the accuracy of these alogrhythms is conditional, depending on how the house being tested compares with the homes used to develop the conversion equation. In this regard, the procedure published in the **ASHRAE Handbook of Fundamentals** is considered the most credible (refer to the section on ventilation and infiltration).

Solar Loads Associated With Glass
In **Manual J**, solar gains are ignored in the heating calculation. This practice produces a conservative estimate of the load associated with an extended period of heavy day-time cloud cover. In the case of the cooling load, the tabulated data provides an estimate of the combined load (solar and conduction) associated with glass, by direction-of-exposure. In this regard, the effects of thermal mass and solar heating are averaged for the whole day. This averaging technique simplifies the procedure and simulates a peak-block-load calculation, which is the appropriate way to estimate the design load for the entire home.

Drapes and Blinds
If a window is positioned so that it does not compromise privacy, it may not be equipped with an interior shading device. In this case, there is justification for using the bare-glass data to estimate the cooling load associated with the window. But this philosophy does not apply to windows in general, because they are typically equipped with drapes or blinds. (This can be verified by taking a leisurely drive through a residential area). Therefore, when privacy is an issue, the internally-shaded-glass data should be used to estimate the window load. (This does not underestimate the solar load because the shaded-glass data represents the load associated with an open blind or a partially drawn drape of roller shade.)

External Shade Screens
The cooling load will be significantly reduced if the windows are equipped with external shade screens. If these devices are installed (or scheduled to be installed), the designer can take credit for the reduction in the solar load, providing the owner understands that they must be permanent fixtures.

Overhangs
Small overhangs, two feet or less, do not significantly reduce the cooling load associated with glass, except for southern exposures; larger overhangs reduce the solar load on all exposures, except for glass that faces north. When applicable, the designer should take credit for the shading provided by an overhang.

Duct Losses and Gains
When all or part of a duct system is installed in an unconditioned space, the exposed duct runs, fittings, boots, and cabinet panels must be tightly sealed and insulated. This attention to detail is necessary because the accuracy of the load estimate is directly related to the designer's ability to model the performance of the duct system. (Duct losses also affect equipment size, equipment capacity, utility demand load, comfort, air quality, and operating cost). In this regard, conduction loads can be quantified with reasonable accuracy, but there is no way to predict the leakage loads unless the duct sealing work conforms to a demanding standard. (Surveys indicate that the leakage rate associated with an unsealed duct system could range from about 10 percent to more than 40 percent of the blower CFM.) Refer to **Manual D**, Sections 12 and 13 for more information on this subject.

Conduction Loads
The structural-component conduction loads caused by the indoor-outdoor temperature difference can be calculated with reasonable accuracy. When making these calculations, the designer should award full credit for documented insulation work. If there is doubt about the construction detail or the quality of the workmanship, a conservative estimate of the effective R-value will suffice.

Mass and Solar Effects — Heating
In the case of the heating loads, no credit is taken for the thermal mass of the structural material or the heating effect of the sun. This practice is consistent with traditional procedures that simulate the worst-case scenario of an extended period of low temperatures and heavy day-time cloud cover.

Mass and Solar Effects — Cooling
As far as the cooling load is concerned, the effects of thermal mass and solar heating cannot be ignored, so they are averaged for exposure and time-of-day. This averaging technique simplifies the procedure and simulates a peak-block-load calculation, which is the appropriate way to estimate the design load for the entire house.

Radiant Barriers
Radiant barriers typically are installed with an intent to reduce the cooling load associated with a ceiling below an attic. The effectiveness of this practice depends on the R-value of the ceiling insulation and on how well the attic is ventilated. If the ceiling is adequately insulated and vented, the benefit will be relatively small. Conversely, if the traditional construction is relatively inefficient, the radiant barrier will produce a significant reduction in the ceiling load. In any case, knowledgeable designers understand that because the benefit is conditional, the effectiveness of a radiant barrier cannot be characterized by an "equivalent R-value." (Refer to Section 4-4 for more information about radiant barriers and equivalent R-values.)

Ventilation Loads

Ventilation will be required if the home is too tight or if mandated by a code or regulation. If ventilation is provided, the increase in the heating load, and the increase in the sensible and latent cooling loads, will depend on how the outdoor air is introduced to the conditioned space. If raw outdoor air is ducted to the return-side of the air distribution system, or if it is drawn or dumped into the conditioned space by a fan, the added load will depend on the difference between the temperature and humidity of the room air and the outdoor air, and on the ventilation CFM value. If the outdoor air passes through an energy reclaim device before it enters the conditioned space, the load on the central equipment will depend on the temperature and humidity difference between the room air and the air that is discharged from the heat reclaim device, and on the ventilation CFM value. (Refer to Section 5-8 more information about heat-recovery ventilation systems.)

Internal Loads

Manual J provides a generic set of values for the internal loads (sensible and latent) generated by occupants and appliances. If an on-site survey produces a more accurate estimate of these loads, they can be substituted for the traditional Manual J values.

Accuracy of the Procedure

When meticulously executed, the Manual J procedure produces conservative estimates for the equipment-sizing loads. The size of the error band is not documented, but some utility-sponsored research indicates that the margin of safety could range from 10 percent to 20 percent, depending on the application. This means that there is no need to hedge on the design conditions, or to derate the thermal efficiency of the architectural features. If excess capacity is desired, it can be provided when the equipment is selected (refer to Manual S for sizing guidelines).

Some designers like to make a "soft load estimate," just to be on the "safe-side." This philosophy is discouraged because it can result in a meaningless load calculation. For example, consider what happens when a designer fudges on the outdoor design temperature, underrates the effectiveness of the insulation, overestimates the infiltration rate, and refuses to take credit for drapes and blinds. This attempt to provide a "cushion" will result in a generous, but undocumented factor of safety — 10 percent here, 15 percent there; by the time a grand total is generated, the load could be overestimated by 30 percent or more. And then, when the equipment is selected, this undocumented error-band might be inflated by another "overall" safety factor.

Zoned Systems and Limited Exposures

If a zoned system is used, or if a dwelling only has one or two exposures, estimates of the peak load associated with the various zones and the central equipment will be required. A procedure for calculating these loads is provided in Manual J. (Refer to Section 3 for more information about zoned systems.)

10-4 Equipment Sizing

A decision to use a specific type of fuel conversion equipment is the primary product of the system selection debate (step 1, Figure 10-1). The minimum size of this equipment will be dictated either by the traditional Manual J load calculation for the entire house (single zone system), or a modified Manual J procedure that estimates the peak block load associated with a collection of rooms (multi-zone system or a dwelling with one or two primary exposures). There also is an upper limit on the size of the equipment, which is dictated by considerations associated with part-load comfort, installed cost, and operating cost.

Undersizing

The obvious problem with significantly undersized equipment is that it will not maintain the desired setpoint when a passing weather system simulates a design condition. However, slightly undersized equipment — by a margin of 10 percent or less — may actually provide more comfort at less cost (see the following discussion of part-load operation). Also, if a multi-zone system is featured, there is an opportunity to abandon the concept of maintaining the indoor design condition in every room of the home for every hour of the year. In this case, the installed capacity could be based on the total load associated with a group of fully conditioned rooms and a group of partially conditioned rooms.

Oversizing

Excessively oversized equipment short-cycles, marginalizes part-load temperature control, creates pockets of stagnate air (unless the blower is operated continuously), degrades humidity control (cooling equipment), requires larger duct runs, increases the installed cost, increases the operating cost, increases the installed load on the utility grid, and causes unnecessary stress on the machinery. Refer to Section 6 of this manual or to Manual S for more information about the limit on excess capacity associated with specific types of fuel conversion equipment and supplemental heating coils.

Even if the deleterious effects on comfort and reliability are discounted, excessive refrigeration-cycle capacity cannot be justified unless the reduction in annual operating cost is large enough to compensate for the incremental increase in installation cost. In this regard, unbiased calculations indicate that there is no economic benefit associated with installing an oversized heat pump (air- or water-source) when a home is located in a mild or hot climate (Seattle or Atlanta, for example). However, if a heat pump is installed in a cold climate (Akron or Minneapolis, for example), a large percentage of the total energy bill will be for heating, so there could be some economic justification for installing a heat pump that has excess cooling capacity. This benefit, as defined by a rate-of-return, will depend on the efficiency of the structure, the equipment performance characteristics, the climate, the utility rate schedule, and if applicable, a utility incentive program. Refer to the "optimum balance point" discussions in **Manual S**, Section 4 (air source heat pump) and Section 5 (water source equipment) for quantitative information on this subject.

Part-Load Performance

Figure 10-4 shows an example of the relationship between the cooling capacity, which depends on the outdoor temperature (air-cooled equipment), and the cooling load, which varies with the outdoor temperature and the solar load. Note that if the equipment is sized so that there is about one-half ton of excess cooling capacity (the design temperature in Atlanta is 92°F and 88°F in Minneapolis), the unused capacity will be about 50 percent of the available capacity when the outdoor temperature is equal to 85°F and about 33 percent at 90°F. As indicated by the bin hour information on the diagram, this piece of equipment will be significantly oversized for all of the bin hours below 90°F, which represent more than 90 percent of the cooling season. As explained in Section 1-6 of this manual, this surplus capacity will translate into diminished humidity control for most of the summer. (A similar diagram drawn for the heating season would show similar excesses at part-load conditions. In this mode, humidity control is not an issue, but as noted above, there are a variety of reasons to minimize excess capacity.)

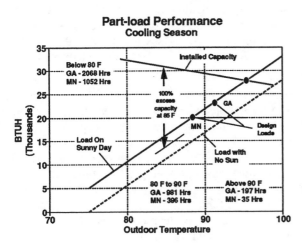

Part-load Performance
Cooling Season

Figure 10-4

Efficiency Rating

When a piece of equipment is selected, the designer's first obligation is to deliver comfort, which means that the machinery must be able to control temperature and humidity during any reasonable operating condition. This test will reduce the list of candidates to a few products that differ in efficiency and price. Since comfort will be assured by any unit

on this list, the designer should select the unit that optimizes the homeowner's capital investment. In this regard, the equipment that has the highest efficiency rating may not produce the most desirable return-on-investment. Therefore, the designer should be prepared to demonstrate that the marginal cost associated with purchasing more efficiency is justified by a useful decrease in the cost of operation. In some cases, incentive programs offered by utilities or lending institutions will be required to make the high efficiency equipment attractive to the homeowner.

Sizing Procedure

The objective of the equipment sizing procedure is to match equipment capacity with the **Manual J** loads for a specific set of operating conditions. In order to satisfy this requirement the designer must produce an accurate load estimate and the equipment manufacturer must provide comprehensive application data. The procedures associated with this work are briefly discussed in Section 6 of this manual and comprehensively demonstrated in **Manual S**.

Variable Speed Equipment

Variable or multi-speed equipment produces more comfort during a fractional load condition, and if a heat pump is featured, speed control can compensate for a mismatch between the design heating load and the design cooling load. For example, if the heating load is large compared to the cooling load, all the refrigeration cycle capacity can be used during the heating cycle, and cooling can be provided at reduced capacity. This way, summer humidity control need not be traded for a lower balance point. Refer to **Manual S**, Section 7 for more information on this subject.

Blower Performance and Ancillary Devices

Primary equipment is initially selected on the basis of its ability to deliver the desired heating or cooling capacity, and on its efficiency; but unless the equipment is completely self-contained, other performance characteristics must be considered. For example, if forced air equipment is designed to interface with a duct system, the blower must be able to overcome the resistance created by the straight runs, fittings and air-side devices. If the blower cannot satisfy this requirement, or if the pressure drop across an external device is too large, the designer must find an alternative product. More information about this issue is provided in Section 6 of this manual, or refer to **Manual S**.

Blower and Device Pressure Drop Data

In addition to capacity and efficiency data, manufacturers publish blower performance data and other tables that document the pressure drop across ancillary air-side devices not tested with the blower (usually a coil, filter, or humidifier). This information should be collected because it will be required for the duct sizing calculations.

Controls and Accessories

The final step in the equipment selection procedure is to make sure that all of the necessary safety controls, operating controls, and accessories are installed with the equipment. In this regard, homeowners do not know much about crankcase heaters, suction-line accumulators, refrigerant-line filters, liquid-line dryers, service-gage ports, liquid-line solenoids, discharge mufflers, hard-start kits, high- and low-pressure switches, anti-short-cycle relays, motor overload protection, and defrost controls. Contractors should explain why these devices are needed and how they affect the cost of the product.

10-5 Air Distribution

Figure 10-5 illustrates the energy and mass flows associated with the forced air cooling process. Notice that the flow of supply air enters the room at a lower temperature, say 55°F, and leaves the room at 75°F (determined by the thermostat setpoint), which means that the air absorbs heat as it flows through the room. In fact, if the indoor setpoint is to be maintained at 75°F, the flow must flush 3,000 BTUH of sensible heat from the room (exactly).

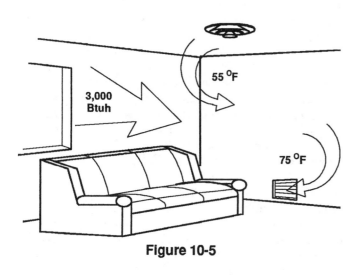

Figure 10-5

Since one would not expect 100 CFM of supply air to have the same flushing capability as a 1,000 CFM flow, there must be a relationship between the sensible load, the entering temperature, the leaving temperature, and the rate of air flow. This relationship is known as the **sensible heat equation**, and Figure 10-6 shows that when this equation is applied to the previous example, the air-side flow rate must be equal to 136 CFM. This means that no other CFM value will balance the heat removing capability of the supply air with the load. If the flow of supply air is throttled, the indoor temperature will increase and the sensible load will decrease (because of the lower indoor-outdoor temperature difference) and a new balance will be established; or the indoor temperature will drop and the cooling load will increase if the flow rate is increased. (The sensible heat equation also governs the heating process, but in this case, heat flows out of the home, and the supply temperature is warmer than the return temperature.)

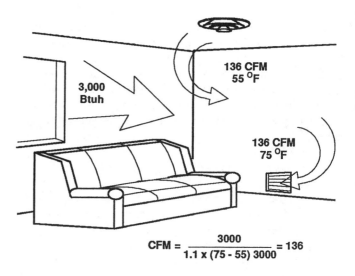

$$CFM = \frac{3000}{1.1 \times (75 - 55)\,3000} = 136$$

Figure 10-6

Notice that the sensible heat equation makes absolutely no reference to the size of the conditioned space. It will require 136 CFM to control the temperature in this example whether the room has 300 or 600 Sq.Ft. of floor area, or any other floor area value. This means that rules-of-thumb regarding supply CFM per-square-foot of floor area, or minimum air-turnover rates, should not be used to design a comfort-conditioning system. However, the air motion within the conditioned space is still a fundamental concern because a moderate amount of air movement is necessary for comfort. This quandary can be resolved by installing air distribution hardware that is compatible with the calculated (load-based) airflow values, the location of the duct runs, the size and shape of the rooms, and the construction details associated with the rooms.

Forced-air equipment manufacturers publish performance data that correlate equipment output capacity with the condition of the entering air and the blower CFM (see Figure 6-2). But, they do not publish a value for the corresponding leaving air temperature, probably because this parameter is a dependent variable. (If the sensible capacity, CFM, and entering temperature are known, a value for the leaving air temperature can be obtained by using the sensible heat equation.)

In other words, when packaged equipment is selected, a value for the design CFM is automatically established when the equipment capacity (heating, or sensible and latent) is matched with the design load and the entering conditions — as demonstrated in **Manual S**. Therefore, the sensible heat equation is not normally used to design a residential air distribution system. Instead, room CFM values are based on the ratio of the sensible room load and the total sensible load on the central equipment (as explained in **Manual D**).

Supply Outlets Have Mixing Power
If a supply outlet is properly sized, a jet of air will be projected away from the outlet at a velocity ranging between 500 and 700 feet per minute (FPM); which means that the discharge will have a useful amount of momentum and kinetic energy. If the outlet vectors this discharge parallel to a flat surface (ceiling or wall), most of the momentum will be transferred to the air in the room, causing it to slowly roll around the room; while some of the jet energy is used to overcome the frictional forces that work to retard the rolling action. In this regard, laboratory tests conducted by hardware manufacturers show that 1,500 to 2,000 CFM of room air will be put into motion when 100 CFM of supply air is discharged from a properly sized outlet. Of course, the larger mass of air will be moving very slowly, at about 25 FPM, which is exactly what is needed to make the occupant feel comfortable. (Air moving at less than 15 feet-per-minute feels stagnate and air moving at 50 feet per minute is considered to be a draft.)

Supply Outlet Arrangements
Perimeter outlets installed along an outside wall, on or near the floor, and preferably under a window, provide the best heating performance and adequate cooling performance. This arrangement is especially desirable for heating rooms that have a slab floor or a floor that covers an unheated space. The discharge air should blow straight up the wall — never out into the occupied zone — and the outlet should be sized so that the primary jet of air gently bumps the ceiling at 50 to 100 FPM. (In other words, the outlet should be sized so that the "throw" is compatible with the ceiling height.)

Ceiling outlets provide the best cooling performance, and they can be used for heating rooms that cover a heated space, but this arrangement is not desirable for heating rooms that have a slab floor or a floor that covers an unheated space. This type of outlet can be installed in the center of the room, or it can be off-set, in order to match the throw with the distance to an outside wall. In any case, the supply air should be discharged parallel to the ceiling — never down into the occupied zone — and the outlet should be sized so that the projected jet of air gently bumps the one or more walls at 50 to 100 FPM. (Manufacturers produce hardware that discharges the supply air in a one-, two-, three- and four-way pattern or a circular pattern.)

High inside-wall outlets offer acceptable cooling performance, providing that the jet of supply air does not drop into the occupied zone before it crosses the room. During the heating season, this type of outlet is compatible with rooms that cover a heated space, but this arrangement is not desirable for heating rooms that have a slab floor or a floor that covers an unheated space. As implied above, the discharge air should move parallel to the ceiling — never down into the occupied zone — and the outlet should be sized so that the projected jet of air gently bumps or just falls short of the opposite wall.

Outlets installed in the ceiling, close to the outside wall, can be used to project supply air down along the wall during the heating season. This arrangement provides an alternative to a below-the-slab duct system. In theory, the heating performance should be better than a conventional ceiling or high-inside-wall outlet. However, if the outlet is sized to throw warm supply air all the way to the floor, a drafty condition could be created in the area below the outlet; and if the outlet is sized to avoid this problem, the floor might not be adequately heated. (This problem is acerbated by a large difference between the temperature of the supply air and the room air, and moderated by a small differential.) Also note that unacceptable performance can be expected during the cooling season unless the outlet is equipped with vanes that can be adjusted to discharge the air in a direction parallel to the ceiling.

If the basement has living areas, maximum comfort will be provided at the floor level if the supply outlets are installed at a low-outside-wall position. If this is not practical, baseboard heat could be used to supplement the central system.

Oversized outlets contribute to the classical stratification problem associated with multi-level homes (upper level too hot, lower level too cold). For example, if a second floor ceiling outlet is too large, cool supply air will just drop to the floor, cascade down the stairway or spill off a balcony, and accumulate on the floor at the lower level. In other words, some or all of the supply air conditioning potential will be wasted if the outlet does not thoroughly blend the supply air with the room air.

Ceiling or side-wall outlets installed in great rooms or double-high entrance areas can be used for cooling, but they will not be effective during the heating season. And, the performance of floor outlets is only marginally better, because the supply air just floats to the ceiling after it is projected into the occupied zone. With this type of architecture, an improvement in heating performance might be provided by a downward discharging outlet located on an outside wall (about eight feet high) or under a balcony. But, if this strategy is adopted, these outlets will cause problems during the cooling season unless they are dampered shut, or adjusted to vector the discharge in a horizontal or slightly inclined direction.

Since supply air outlets are marginally compatible with double-high architecture, the designer may decide to adopt one of the strategies listed below. In this regard, the cost of installing the supplemental equipment must be balanced against a comfort benefit that is not easily quantified.

- A ceiling fan can be used to mix stratified air with the air in the occupied zone.

- A ducted, fan-powered recirculating system can be used to capture air that has stratified near the ceiling and to reintroduce it to the occupied zone.

- Baseboard heat can be used to supplement a central system that serves a high-bay area.

• A below-the-floor radiant heating system can be used to supplement the comfort conditioning capability provided by the central system that serves the high-bay area.

Returns

An inadequate return system can restrict system air flow, degrade equipment performance, diminish comfort, cause unpleasant variations in room-to-room temperatures, and increase infiltration in some parts of the home. These problems can be avoided by making sure that there is a low resistance return path for every conditioned space. This means that a central return is compatible only with homes that have an open floor plan. If an interior door can be used to isolate a room from a central return, the room must be equipped with its own return; or it must be connected to a central return by a transfer grille or a short transfer duct. (Door undercuts are not recommended because they must be two to three inches high to be effective.)

The designer also must make sure that the size of a return is matched to the flow associated with the return. This requirement will automatically be satisfied if the return is sized so that the face velocity is 400 FPM or less.

Contrary to conventional wisdom, returns do not have a significant affect on room air motion or the size of a stagnate air pocket. In other words, a low return will not "pull" warm air down to the floor during the heating season, nor will a high return "lift" cooler air up toward the ceiling during the cooling season. This means that returns can be installed in positions that are compatible with the location of the equipment, the location of the primary return duct, and the homeowner's budget.

> A return draws air from all directions, so there is no vectoring power associated with a return until the air is very close to the face of the return. For example, think of the flow through a set of concentric hemispheres. The velocity at the outermost surface will be low because the flow area is relatively large. As the journey continues, the surface areas associated with the hemispheres decrease exponentially, with a corresponding increase in velocity; but even at a distance of a few feet, the air will still be moving at a leisurely pace. Finally, when the air is within inches of the return, it will accelerate to a velocity of about 400 FPM. When this ambivalent exodus is compared to the frantic mixing action provided by a jet of supply air, there can be no question about what forces control the air movement within the conditioned space.

Noise

Supply outlets and returns should not generate objectionable noise. This problem can be generally avoided if the velocity associated with supply outlets and returns is limited to 700 FPM and 400 FPM respectively. (These velocity-based guide-lines are, by necessity, conservative. A higher velocity limit can be used if a sound meter test confirms that the noise level is acceptable.) Also, refer to noise generation data (if available), provided by the supply outlet manufacturer. (More information about noise ratings can be found in **Manual T**, Section 9.)

Pressure Drop

Information about the pressure drop across an air distribution device is required for the duct sizing calculations. This information is normally provided by the manufacturer, but the pressure drop across grilles and diffusers normally does not exceed 0.03 inches-water-column (IWC).

Manufacturer's Performance Data

Manufacturer's performance data is required for outlet selection. As indicated by Figure 10-7, tables are used to correlate flow rates (CFM) with throw, velocity, pressure drop, and in some cases, spread. Refer to **Manual T** for a comprehensive discussion of the performance characteristics of various types of air distribution devices and the corresponding performance data.

Floor Diffuser							
Face Velocity		400	500	600	700	800	900
Pressure Loss		.010	.016	.022	.031	.040	.050
2-1/4 x 10	CFM	35	45	50	60	70	75
	Spread	4	5	6	7	8	9
	Throw	3	4.5	5	6	7	8
2-1/4 x 12	CFM	40	50	60	70	80	90
	Spread	4.5	5.5	7	8	9	10
	Throw	4	4.5	5.5	6.5	7	8
2-1/4 x 14	CFM	40	60	70	80	90	105
	Spread	4.5	6	7	8	9.5	10.5
	Throw	4	5	6	7	8	9
4 x 10	CFM	50	70	85	100	120	135
	Spread	4.5	6	8	10	11	12.5
	Throw	4	5	6.5	7.5	9	10
4 x 12	CFM	80	100	120	140	160	175
	Spread	6.5	8	9.5	11.5	13	14.5
	Throw	5.5	7	8	9.5	11	12
4 x 14	CFM	90	115	140	160	185	205
	Spread	7	8.5	10	12	13.5	15.5
	Throw	5.5	7	8.5	10	11.5	12.5
Velocity at end of throw = 50 FPM							

Figure 10-7

10-6 Distribution System Design

The information generated during the first four steps of the design procedure provides the basis for the conceptual decisions and sizing calculations associated with the distribution system design work. At this point the designer knows where the primary equipment will be installed, where the supplies and returns will be located, the appropriate location for the duct system (refer to **Manual D**, Section 1), the Blower CFM, and the CFM associated with each supply outlet or return. At this point, the designer will select a particular type of distribution system (normally, a trunk and branch arrangement or radial layout), and a fabrication material (sheet metal, duct board, or a flexible wire helix product). After these decisions are made, the design work will consist of selecting routes for the duct runs, selecting fittings, sizing the trunk sections (when applicable), and sizing branch runouts.

Routing

Duct runs should be routed as directly as possible to minimize the number of fittings along the path. Unless the system is installed in the attic or a marginally accessible crawl space, this work usually requires a compromise among air-side efficiency, encroachment on usable living space, and conflicts with work installed by other trades. In any case, the practice of using panned air-ways or framed chases to create a unobtrusive conduit is not recommended because of the difficulty associated with sealing the path.

Critical Circulation Path

The blower is normally connected to the conditioned space by multiple supply runs. These runs are in parallel — like parallel resistors in an electric circuit — and if there is more than one return run, there will be a second set of parallel conduits on this side of the system. As far as flow resistance is concerned, the performance of any collection of parallel conduits depend on the total effective length of the longest supply run and the total effective length of the longest return run. This particular conduit pair is referred to as the "critical circulation path." (Refer to **Manual D** for instruction and examples pertaining to the use of the Effective Length Calculation Sheet.)

Fittings

The performance of the fittings that provide the interface among the duct runs; the equipment; the air distribution devices; and the fittings that are used to turn the flow, to create a branch flow, to merge two flows, or to change the shape of the conduit are characterized by equivalent length values. An inspection of **Manual D**, Appendix 3 indicates that there can be a wide range of equivalent length values associated with a particular class of fitting. This variety gives the designer considerable control over the effective length of the critical circulation path. Therefore, when the designer attempts to match the resistance of the duct system to the capability of the blower, fitting efficiency can mean the difference between success and failure.

Available Pressure

The duct system must be designed to work with the blower that is packaged with the heating and cooling equipment. In this regard, the first task is to determine how much static pressure will be available to move the air through the fittings and the straight sections of duct. This pressure will be equal to a value listed in the manufacturer's blower performance table, discounted for the total pressure drop associated with components that were not installed when the blower was

tested. (Components which were installed during the blower test are usually listed in footnotes located below the blower table. The pressure losses associated with accessory devices are usually compiled in supplementary tables.)

For example, Figure 10-8 indicates that when operating at medium speed, the blower can move 1,250 CFM against an external resistance of 0.49 IWC. In this case, the available pressure does not have to be discounted for the pressure drops associated with a wet coil or a standard filter because these components were in place when the blower was tested. However, an electric heating coil (heat pump application) is an accessory device, as indicated by the note below the blower table. This means that the designer must return to the engineering data to determine the pressure drop across the resistance coil; as indicated by Figure 10-9, 0.14 IWC must be subtracted from the blower pressure value. In addition, the available pressure must be adjusted for the pressure drops associated with every ancillary air-side device that is in the critical circulation path. (Deductions are always required for the supply outlet, the return grille, and the branch balancing damper. Other reductions may be associated with an accessory filter, a humidifier, or any other type of pressure dissipating device.)

Blower Performance			
CFM	External Static Pressure IWC		
	High	Med	Low
1200			0.45
1250		0.49	0.30
1300		0.37	0.08
1350		0.25	
1400	0.62	0.14	
1450	0.55	0.04	
1500	0.47		
1550	0.39		
1600	0.31		

Tested with wet coil and filter in place. Subtract pressure drop associated with resistance heating coil.

Figure 10-8

Electric Resistance Coil Pressure Drop	
CFM	IWC
1100	0.09
1200	0.13
1250	0.14
1300	0.16
1400	0.19

Figure 10-9

Variable Speed Blowers

Variable speed blowers can operate over a wide range of flow and pressure conditions. For example, by adjusting the rotational speed of the wheel, one of these devices could be used to deliver 1,250 CFM against system resistances that range from less than 0.20 IWC to more than 0.80 IWC. This flexibility can compensate for poor design work, but it should not be used as a rationale for circumventing the air-side design process. (Refer to **Manual D**, Section 5 for information about variable speed blowers.)

Friction Rate Sizing Value

Since duct friction charts and duct sizing slide rules correlate duct sizes and flow rates (CFM) with a friction rate (pressure drop in IWC per 100 feet of duct, or F/100 value), the available pressure value cannot be used to size duct runs. But, it is one of the two factors that determine what friction rate will be used to size the duct runs. The other factor is the total effective length of the critical circulation path. The following equation shows how these two factors are used to generate the *design friction rate value*. (Refer to **Manual D** for instruction and examples pertaining to the use of the Friction Rate Worksheet.)

$$F/100 = \frac{\textbf{Available Pressure x 100}}{\textbf{Total Length of Critical Circulation Path}}$$

Design CFM Values

The CFM that must be delivered to a room depends on how the size of the room load compares to the total load on the central equipment. For example, if the room load is equal to 15 percent of the total load, the room should receive 15 percent of the blower CFM. Of course there may be two loads (heating and cooling), in which case two CFM values would be generated, with the larger value dictating the design condition.

Note that the room CFM value is not always equal to the branch-duct flow rate because multiple outlets are used when a room requires a substantial amount of air (more than 140 CFM). In this case, the CFM associated with each branch run will depend on the air distribution plan.

On the return side, the flow rate associated with a fixture and its attendant branch run is dictated by the supply outlets that can logically be grouped with the return. This means that the designer must identify open areas, areas that can be coupled by a transfer duct or grille, and isolated rooms. For example, all of the supply CFM would be routed through a large, central return if the home features a completely open floor plan. Or an area return might serve two or more rooms, or a proprietary return might be installed in an isolated room.

The flow rate through a trunk duct is cumulative as the point of observation moves toward the blower. This means that the CFM associated with any section of supply trunk duct is equal to the sum of the downstream branch flows, and the CFM associated with any section of return trunk is equal to the sum of the upstream branch flows.

Noise and Velocity

Initially, the size of a duct run is determined by the flow rate and the design friction rate value. This practice produces a design that will be compatible with the blower's capability, and if generated noise was not a consideration, the sizing exercise would be complete. However, objectionable noise can be generated in a duct run if the velocity of the flow is too high. When this is the case, the size of the offending run will have to be increased. In this regard, the velocity limit (refer to **Manual D**, Table 3-1) depends on the location of the duct with respect to the blower (supply or return), the type of run (trunk or branch), and the duct material.

Final Design

The final design work begins with a matrix of round sizes that are compatible with the friction rate design value and the CFM associated with the duct sections. When necessary, the duct slide rule can be used to convert round sizes into rectangular sizes that have an equivalent friction rate. Then the duct slide rule can be used to check the velocities associated with each size (use the front side of the slide rule for round ducts and the back side for rectangular ducts). If a velocity is too high, the duct section is resized to comply with the recommended limit. (Refer to **Manual D** for instruction and examples pertaining to the use of the Duct Sizing Worksheet.)

Balancing Dampers

Balancing dampers must be installed in each runout duct. This is necessary because the flow rate associated with a run that is not in the critical path, or any run that has been resized to satisfy a velocity limit, will be excessive unless the pressure drop is adjusted to match the critical-path pressure drop. In this regard, registers are not suitable balancing devices because they generate noise in the hard-throttle position.

10-7 Air-Side Balancing

Air-side balancing is not a design task, but since this work mediates conflicts between installed system performance and design objectives, it falls within the purview of the system designer. This means that balancing is required even when the design calculations are comprehensive and the installation procedures are perfect. (All controls must be checked and adjusted, a fan-speed adjustment should be made when required, and all the balancing dampers should be adjusted and locked.)

Section 11
Energy and Operating Cost Calculations

Computer software that performs energy and operating cost (op-cost) calculations can be used as a design tool, to demonstrate code compliance, to improve the terms of a mortgage (monthly utility bills are factored into projected cash flows), or to generate information for a sales presentation. However, the utility of a computerized calculation will be marginalized if the program is difficult to use or if it has limited application. In this regard, the calculation tool must be capable of modeling a wide range of structural features and mechanical devices, and the calculations must be based on local weather patterns. The software also must provide outputs (energy loads and utility costs) that are reasonably accurate.

11-1 Utility of Economic Calculations

During the design process, energy and op-cost calculations can be used to evaluate the cost-benefit relationship associated with a system concept, an equipment sizing strategy, an additional increment of equipment efficiency, an improvement in the efficiency of the distribution system, or an improvement in the thermal performance of a structural component. This process begins with the first visit with the homeowner (or builder), and ends when the homeowner and the local building inspector (if applicable) approve the design concept.

System Design Tool
For residential applications, the "system" includes the components of the building envelope; the lighting, appliances, and fireplaces; the family members; the HVAC equipment, distribution hardware, and vents; and the kitchen and bath exhaust equipment. In this regard, the HVAC contractor usually arrives on the scene after the structural design has been finalized, so from the comfort contractor's perspective, the design work pertains to system selection, equipment sizing decisions, and control strategies.

For example, an op-cost calculation could be used to estimate the cost-benefit associated with installing a multi-zone system or an earth-coupled, water-loop system. Or, an economic calculation could be used to compare the costs of installing and operating a fossil fuel furnace with a water-source heat pump, or an air-source heat pump. Or, an op-cost calculation could be used to justify an equipment-sizing strategy (refer to **Manual S** for information about heat pump sizing limits as they pertain to the relationship between first cost, op-cost, thermal balance point, and excess cooling capacity). An economic analysis also can be used to evaluate the wisdom of using a set-up—set-back control strategy (scheduled set-point changes do not always translate into reduced fuel costs, or the savings may be less than expected.)

Occasionally, the comfort-conditioning contractor may be required to investigate the relationships between the efficiency and cost of the structure, and the efficiency and cost of the HVAC system. For example, if a home has a generous amount of glass area, replacing single-pane windows with high-efficiency windows can reduce the load on the HVAC equipment by more than 50 percent, which could translate into an approximately equal reduction in the annual utility bill (depending on the glazing plan and the location of the home site).

Robust computer models also can generate detailed information about important design parameters. Some of the items of interest to a designer might include the thermal balance point (heat pump equipment), the economic balance point (dual-fuel application), auxiliary heating requirements (heat pump equipment), the cooling-heating balance point of the structure (no load at the change-over temperature), and the operational coil sensible heat ratio (cooling equipment). The output also might provide a list of performance-index values that can be used for error trapping. (Items of interest might include information about the required equipment capacity per-square-foot of floor area for cooling and heating, the CFM per-square-foot of floor area, a CFM-per-ton value, the energy consumption per-square-foot of floor area for cooling and heating, and site-specific values for full-load cooling and heating hours.)

Compliance With Standards and Codes
A comprehensive computer model provides a tool that can be used to demonstrate compliance with energy codes on a "system performance" basis. (The system performance approach is desirable because it provides the maximum amount of design freedom.) When used in this capacity, compliance is demonstrated by showing that the annual energy budget for the proposed house is smaller than the annual energy budget for a similar house that meets a prescriptive standard.

Useful Sales and Marketing Tool
The output of a thoughtfully designed computer model should consist of text, tables, and graphics that would be suitable for making presentations to prospective customers. In this regard, a payback analysis could be used to demonstrate whether a high-efficiency system, or a piece of high-efficiency equipment is cost effective. The software tool also could be used to demonstrate the benefit associated with upgrading the efficiency of an air distribution system or a structural component.

11-2 The Calculations Should Be Site-Specific

With the power provided by the typical desktop or laptop computer, trade-off studies and economic evaluations should require a minimum number of extrapolations and assumptions. This means that calculation tools that rely on aggregated weather data (heating degree days and cooling degree hours, or heating and cooling load hours), generalized efficiency descriptors (AFUE, HSPF, and SEER), or a virtual reality created from a regression analysis of a statistical data base, can be replaced with calculations that analyze a specific structure, in a particular city or town, using performance models for a specific type of conditioning equipment and distribution system. In other words, assumptions about weather patterns, envelope performance details, equipment efficiency, distribution system efficiency, and energy costs should not be overtly or covertly associated with the calculation procedure.

11-3 Ease of Use

Theorists and software engineers are responsible for providing a calculation tool that is easy to use. All the inputs should be based on information commonly found in a contractor's office, or it must be supplied with the calculation tool (in a format as transparent to the user as possible). As indicated by Figure 11-1, most of the required input is generated by the **Manual J** load calculation procedure and **Manual S** equipment sizing procedure. Therefore, if the user can provide some fuel-cost information, and select internal load and thermostat setpoint schedules from a menu, the energy and op-

Input for Computerized Energy and Op-cost Calculations					
	Weather Model	**Envelope Model**	**Equipment Model**	**Control Strategy**	**Utility Charges**
Data Required	Hourly, or bin data; dry-bulb and coincident wet-bulb temperatures; peak solar load by azimuth and latitude; percent sunshine; and wind velocity	Information about geometry and type of construction (walls, floors, ceilings, roofs, windows, and doors); tightness of structure; ventilation requirement	Information about equipment capacity, efficiency, and power requirements; information about duct system efficiency	Information about zones, heating and cooling setpoints, set-up and set-back schedules, timed defrost setting, supplemental heat lock-out, etc.	Information about the unit cost of fuel, utility rate structures, and demand charges (heating and cooling)
Source	Internal data base	Information extracted from building plans or field survey	Performance data published by equipment manufacturers	Parameters extracted from design concept and load calculation input	Local utilities and suppliers
Primary Input Required	Select city from menu	Enter **Manual J** data in menu-driven screens; select internal load schedule and setpoint schedules from menus	Enter operating points extracted from published application data in menu-driven screens (or extracted from an internal database)	Select control schedules and strategies from menus; enter setpoints on screens (some setpoints extracted from load calculation data)	Enter data on screens (to be stored internally for update, as required)
Supplemental Information	None	Data to model thermal mass can be generated internally (some parameters might be selected from a menu)	Information about duct losses, defrost control, supplemental heating capacity, well depth, water pump, entering water temperature etc.	Continuous fan option; dual-fuel option; multiple or variable speed option	Notes about incentive programs and rebates
Comments	Bin data should be segmented by month and hour-of-day groups	Dynamic model of structure generated by the tool interacts with the weather, equipment, control, and fuel-cost models	Dynamic model of equipment generated by the tool interacts with the weather, envelope, control, and fuel-cost models	Control cycle information integrated with the equipment performance model's response to envelope loads	A collection of input screens should accommodate all of the common billing structures

Figure 11-1

cost calculations become an automated by-product of the standard design procedure. (If an equipment performance data base is provided within the tool, the **Manual S** equipment sizing procedure also could be automated.)

11-4 Range of Application

Figure 11-2 indicates that a basic calculation tool should be able to analyze the most common types of residential structures and single-zone systems; and that a more sophisticated tool would be used to generate energy-load and op-cost estimates for the various types of multi-zone systems. In either case, the associated thermal-mass models must be compatible with the desired control strategy (constant set-point or scheduled setpoint), and additional modeling capability is required to accommodate systems that feature multi-speed and variable speed equipment.

11-5 Compatibility With Codes and Rating Systems

Two op-cost calculations are required to demonstrate compliance with an energy code, home efficiency rating system, or financial incentive program on a "system performance" basis — one for the proposed design, and an equivalent calculation for a geometrically similar structure that conforms to a prescriptive menu. In this regard, it is not difficult to design a calculation tool that can transparently convert the data that describes the proposed structure into a model of the codified benchmark structure. This way, the work associated with creating and analyzing the prescriptive design is effortlessly performed by the computer.

Calculation tools also should automate the work associated with using less sophisticated compliance paths. For example, the tool should be able to compare the data that describes the proposed design with a prescriptive standard, and generate an

Suggested Levels of Analysis Capability for Calculation Tools			
	Level 1	**Level 2**	**Level 3**
Structure	Single zone, no unusual architectural features	Single zone, multi-zone, no unusual architectural features	Levels 1 and 2, large amounts of glass, passive solar features, indoor pools and hot tubs
Primary Equipment	Boiler, Furnace, Air conditioner, Air-source heat pump, Water-source heat pump, Furnace and heat pump, Electric resistance coils	Level 1 equipment, Multiple single-zone units, Variable Air Volume Systems, ductless, split-coil systems	Levels 1 and 2, Thermal storage equipment, Solar panel systems
Fuels	Electric, gas, oil, and LP	Electric, gas, oil, and LP	Electric, gas, oil, and LP
Sources and sinks	Atmosphere, well water and earth-coupled piping loop	Atmosphere, well water and earth-coupled piping loop	Atmosphere, well water and earth-coupled piping loop, masonry, rocks, water tank
Speed Control	Single speed, multi-speed	Single speed, multi-speed, and variable speed	Single speed, multi-speed, and variable speed
Secondary Equipment	Supplemental electric coils, Blowers and fans, Water pumps, Crankcase heaters, Ignition devices, pilot lights	Supplemental electric coils, Blowers and fans, Water pumps, Crankcase heaters, Ignition devices, pilot lights	Level 2 plus system-specific devices
Controls	Thermostats, Set-back and set-up, Defrost cycle, Supplemental heat, Crankcase heater	Room thermostats, Central control panel, Set-back and set-up, Defrost cycle, Supplemental heat, Crankcase heater	Level 2 plus system-specific devices
Comments	Applies to large majority of housing stock	Consider inter-zonal heat transfer; increasing percentage of housing stock	Enormous amount of modeling required for insignificant percentage of housing stock

Figure 11-2

itemized pass-fail report. In addition, the tool should be able to generate screens that can be used to match the overall thermal envelope conductance (as described by the UA value) associated with the proposed design, with the overall thermal envelope conductance of a geometrically similar structure that conforms to a prescriptive menu. (Equivalence is achieved by adjusting the U-values and R-values of the proposed design's structural components.)

11-6 Reports

The calculation tool should be able to generate a variety of reports that combine text, tables, and graphic exhibits. These formatted presentations should be able to summarize the equipment selection and sizing calculations, the efficiency and op-cost calculations, and the information required to demonstrate code compliance. In addition, the software tool should be able to generate optional reports that provide detail about any aspect of the calculation procedures. (This feature is useful for error checking, what-if calculations, and trouble shooting.)

Design Calculation Summaries
A summary of the design calculations should include information about the **Manual J** heating and cooling loads for the structure, and a breakdown of the loads associated with the structural components. The report also should provide information about the required equipment capacity (at the applied design condition) and related subjects (the design-condition coil sensible heat ratio, the thermal and economic balance points, the supplemental heat requirement, system air-flow rates, and water-flow rates, for example).

Efficiency and Op-cost Summaries
An energy and op-cost summary should include information about the annual heating and cooling energy required for the thermal envelope (the delivered energy requirement) and a breakdown, by fuel type and device, of the heating and cooling energy consumed by the HVAC equipment (the purchased energy requirement). The report also should include information about the cost of the fuel and electrical energy consumed by the comfort conditioning equipment, by season, fuel type, and device. Additional useful information might include calculated, on-site efficiency descriptor values (applied SEER, AFUE, HSPF), and energy budgets for the thermal envelope and the comfort conditioning system on a per-square-foot basis.

Code-Compliance Summaries
The format of a code-compliance summary would depend on the code, the rating system, or the incentive program that has legal standing or market acceptance at the home site. In this regard, the calculation tool should generate a standard set of forms that demonstrate compliance with national and regional documents (CABO-MEC, or the Southern Building Code, for example); and the tool should have a post-processing feature that allow the creation of custom forms (or for modifying a library of forms) that could be used to demonstrate compliance with a state or local code.

11-7 Output Graphics

Graphical exhibits are valuable tools for summarizing the performance of the structure and equipment, for presenting this information to a homeowner or code official, and for checking errors. A partial list of useful charts and diagrams is provided below.

- Bar charts can be used to show the bin-hour distributions that represent the basis for the weather model.

- Bar charts can be used to compare the energy requirements and utility charges associated with various design scenarios and trade-off options.

- Pie charts can be used to show what percentage of the design heating and cooling load is associated with the various structural components, the structural leakage, the ventilation requirement, and the duct losses.

- Load lines (heating load or cooling load plotted against outdoor air temperature) can be used to summarize the performance of the structure for a full range of weather conditions and time-of-day scenarios. (Detail can be provided by displaying the lines that represent the transmission, solar, infiltration, ventilation, internal, and duct loads under the line that represents the combined load.)

- Performance maps for refrigeration-cycle equipment (air- or water-source) can be superimposed on load-line plots to produce balance-point diagrams for heating and cooling. In this regard, separate diagrams can be created for the design heating condition (no credit for solar and internal gains), the design cooling condition (full impact of solar and internal gains included), and intermediate operating conditions (simultaneous presentation of load lines that are biased and unbiased for the effects of solar gains and internal loads).

- For refrigeration cycle equipment, a plot of the system coefficient of performance, or the energy efficiency ratio (against outdoor temperature), can be used to show how system efficiency is affected by operating-cycle losses over a complete range of load conditions. In this regard, dynamic efficiency depends on the energy consumed by the primary equipment, the supplemental heating device (if applicable), the penalties associated with cycling equipment on and off, and the inefficiencies associated with the operation of ancillary devices (a defrost control or crankcase heater, for example.)

- A plot of unit energy cost against outdoor air temperature can be used to evaluate the economic balance points associated with a dual-fuel system.

11-8 Simulation Capability

The capability of any calculation tool depends on the amount of detail that can be addressed by the user, and on the accuracy of the models used to simulate the interaction between the climate, the structure, the equipment, the occupants, the controls, and the utility's pricing schedule. In this regard, software products should include documentation describing the models that can be controlled by user input. This literature also should summarize the models that cannot be manipulated by the user. In addition, the documentation should provide a description of the methods that were used to simulate an effect, a behavior, or a device. Some of the issues associated with building a set of comprehensive models are discussed here.

Weather Module

The weather model should account for the effects of dry-bulb temperature, wet-bulb temperature, solar intensity, latitude, cloud cover, and wind velocity on a hour-of-day basis and on a monthly or seasonal basis. In this regard, maximum detail is provided by hourly-simulations, but adequate detail can be obtained by segmenting bin-hour data.

Structural Module

The thermal envelope model should simulate the sensible loads (transmission, solar, infiltration, ventilation, and internal) and the latent loads (infiltration, ventilation, and internal) for the entire range of applicable weather conditions, thermostat set-point schedules, and internal load schedules. This model also should account for the effect of thermal mass as it applies to solar gains, set-up transients, and set-back transients. In addition, the cooling-load model should account for moisture-load transients associated with set-up schedules. If zoned systems are modeled, the structural model should account for the effect of inter-zonal heat transfers.

Equipment Module

The equipment model should simulate the output capacity and the input power requirement for the entire range of load conditions (heating, sensible cooling, and latent cooling) generated by the interplay between the weather, envelope, and control models. Depending on the type of equipment, these simulations should account for the following collection of interactions, effects, operating modes, and accessories.

- The inefficiency associated with on-off operating cycles, adjusted for the amount of excess capacity associated with design-day operation (application of an appropriate cycling degradation coefficient and over-sizing penalty).

- The effect that duct losses have on the envelope load, the equipment load, and the condition of the air entering the air-side of the fuel conversion equipment.

- The interaction among the cooling coil, the infiltration rate, the ventilation rate, the return-side duct losses (conduction and leakage), the outdoor humidity, and the internal latent load, as it affects the indoor humidity, the entering wet-bulb temperature, the coil sensible heat ratio, and the run-time of the cooling equipment.

- The coil temperature pull-down transient associated with on-off cycling (see Figure 1-7), as it affects the indoor humidity, the entering wet-bulb temperature, the coil sensible heat ratio, and the run-time of the cooling equipment.

- The relationship among the instantaneous load associated with the thermal envelope, the refrigeration-cycle heating capacity, and the supplemental heating capacity, as it pertains to the thermal balance point, and the control scenario (the site-specific supplemental heating energy associated with normal heating-cycle operation).

- The interaction between the weather model (outdoor temperature and humidity) and the defrost-control model, as it pertains to the frequency and duration of reverse-cycle operation, and to the use of back-up heat during these intervals (the defrost cycle energy associated with the site-specific frosting potential at the outdoor coil).

- The relationship between the instantaneous load associated with the thermal envelope, the refrigeration-cycle heating capacity, and the furnace heating capacity, as it pertains to the economic balance point, and the control scenario (the site-specific fossil fuel heating energy requirement associated with a dual fuel system).

- The energy required for intermittent or continuous operation of the indoor fan.

- The re-humidification effect associated with continuous fan operation during the cooling season.

- The energy required for the operation of outdoor fans, adjusted for the effect of the defrost cycle, when necessary. (The outdoor fan is typically de-energized during defrost.)

- The energy consumed by the crank-case heater (depending on the type of control that is used to activate the heater).

- The energy consumed by standing pilot lights.

- The energy consumed by water pumps

- Water piping and refrigerant line losses.

Control Module

The comfort control model should accept input pertaining to the setpoints associated with full comfort conditioning, and it should accommodate input that describes set-up and set-back schedules. (Note that for multi-zone applications, each zone could have an independent set-point schedule. Usually, this modeling is limited to a living-zone, sleeping-zone scenario.) Models also are required to simulate the action of supplemental heating controls (two-stage thermostat, outdoor thermostat, and ramped-recovery thermostat), defrost-cycle controls (demand and time-temperature), crankcase heater controls, and fuel ignition controls.

Fuel-Cost Module

The fuel-cost module should accommodate a variety of rate structures. For example, the unit cost of energy could depend on the season-of-year, the time-of-day, or on the monthly consumption (block rates). In addition, there could be a demand charge if the connected load exceeds a predetermined value; or there could be a special rate for singe-fuel homes, or for systems that include a thermal storage device.

11-9 Single-Value Efficiency Ratings

Single-value efficiency ratings are assigned to gas and oil furnaces (the annual fuel utilization rating, or AFUE), to air-to-air cooling equipment (the seasonal energy efficiency rating, or SEER), and to air-to-air heat pumps (SEER for the cooling mode, and the annual heating season performance factor, or HSPF, for the heating mode). These ratings can be used to compare the efficiency of similar types of fuel conversion equipment when subjected to a very specific set of load and operating conditions. (In this regard, efficiency descriptors are similar to the miles-per-gallon rating that consumers find on the window of a new car.) It is also important to remember that the ratings apply only to single-zone equipment operating to maintain a constant setpoint temperature. There is no recognition of the benefits associated with zoning or set-point scheduling, and there is no allowance for losses associated with an air distribution system.

AFUE

The AFUE rating for gas-fired and oil-fired heating equipment is listed in the *Directory of Certified Furnace and Boiler Efficiency Ratings*, published by the Gas Appliance Manufacturers Associations (GAMA). This rating includes allowances for the effects of on/off cycling, flue losses, and other combustion system factors that affect furnace efficiency. It does not account for the electrical energy consumed by the blower.

SEER

The SEER rating for air conditioners and heat pumps is listed in the *Directory of Certified Unitary Air Conditioners and Heat Pumps* published by the Air Conditioning and Refrigeration Institute (ARI). This rating acknowledges the deleterious effect of on/off cycling, and it includes allowances for the power consumed by the indoor blower (intermittent operation), the outdoor fan, the crank case heater, and the associated controls.

HSPF

The HSPF rating also is listed in the *Directory of Certified Unitary Air Conditioners and Heat Pumps*, published by the ARI. This rating acknowledges the inefficiency associated with on/off cycling, and it includes a defrost cycle penalty. It

also includes an allowance for the power consumed by supplemental heating coils, the indoor blower (intermittent operation), the outdoor fan, the crank case heater, and the controls.

11-10 Comments on Efficiency Descriptors

It is important to understand that published seasonal efficiency ratings (SEER, HSPF, and AFUE) are based on a set of assumptions regarding weather patterns, design heating and cooling loads, installed capacity, distribution losses, ancillary equipment, and control strategy, which means that they have limited application to non-conforming scenarios. A few of the reasons why single-value efficiency descriptors should not be used to estimate the energy load associated with a specific home are presented here.

Effect of Weather

Published efficiency ratings are based on the bin weather data for a specific city (Pittsburgh). This means that if all the aspects of the structural-mechanical design are held constant, the rating could be significantly different in other cities. (In the case of cooling equipment and fossil fuel furnaces, the descriptors completely ignore differences in local weather patterns. In the case of air-source heat pumps, six rating values are correlated with six climate zones, but the zone-4 rating (Pittsburgh) is the only value that is routinely published.)

Equipment Sizing Effects — Cooling

Single-value efficiency ratings are based on an assumed relationship between the design load and the installed capacity. In the case of air-to-air cooling equipment, the published SEER rating is based on a one-to-one load-to-capacity ratio, so a correction is required for applications that involve other load-to-capacity ratios. (The ARI has made an effort to address this issue by publishing an adjustment procedure and a table of correction factors in front of the *Directory of Certified Products*.)

Effect of Latent Loads and Entering Conditions

The SEER rating associated with air-cooled equipment is based on wet-coil tests that are conducted with an 80°F dry-bulb, 67°F wet-bulb condition at the entrance to the indoor coil. However, these values may not be representative of the condition associated with a specific installation because the entering dry- and wet-bulb temperatures depend on the temperature and humidity of the return air (75°F and 50 percent, for example), and on the sensible and latent gains (due to conduction and leakage) associated with the return side of the duct system. (Refer to **Manual S** for examples that show how manufacturer's application data is used to correlate sensible capacity, latent capacity, and input power with the temperature and moisture content of the entering air.)

Equipment Sizing Effects — Heating

In the case of air-source heat pump equipment, the published zone-4 HSPF rating is based on the assumption that the design heating load is equal to the heating capacity of the heat pump when the outdoor air temperature is equal to 47°F. In addition, the procedure used to generate the HSPF value applies a 0.77 "correlation factor" to the assumed heating-load value. (Apparently, the 0.77 correction factor compensates for the load-reducing effect of solar and internal gains. However, there is no reason to expect a generic correction to apply to a specific home.)

The consequences of these assumptions and adjustments can be demonstrated by a balance point diagram. In this regard, Figure 11-3 shows that a minimum amount of supplemental (resistance-coil) heat will be required during cold weather; which means that the published HSPF value will be maximized.

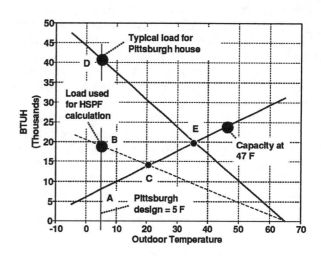

Figure 11-4

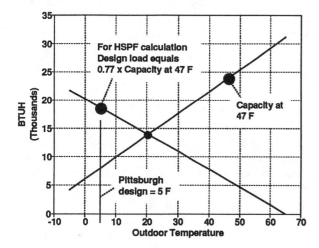

Figure 11-3

The generosity of HSPF calculation procedure is more obvious when Figure 11-3 is compared to a balance point diagram that is representative of a thermally efficient zone-4 home, as illustrated by Figure 11-4. (The example home has 2,000 square feet of floor area, double pane windows, R-19 walls, R-30 ceiling, insulated slab, reasonably tight construction, and a heat pump size compatible with the design cooling load.) Note that more supplemental heat is required for the example home (compare triangle A, B, C with triangle A, D, E), which means that the HSPF value associated with the rating-test scenario will be larger than the HSPF value associated with the example home. (The ARI has attempted to address this issue by publishing an adjustment procedure and a table of correction factors in front of the *Directory of Certified Products*.)

Defrost Penalty
The HSPF rating includes a defrost penalty. However, the corresponding frosting load corresponds to the set of conditions used during the certification test. In practice the defrost penalty associated with a specific heat pump system could be

quite different from the test chamber penalty, because frosting potential depends on local weather conditions as defined by the hourly distribution of the outdoor dry- and equivalent wet-bulb temperatures. In this regard, Figure 11-5 shows how the local weather patterns affect the operation of the defrost control. Note that a large defrost penalty will be associated with a city that has a significant amount of precipitation when the outdoor temperature is in the 30°F to 40°F range. Figure 11-5 also indicates that a small defrost penalty would be associated with a city that has cold, dry winters.

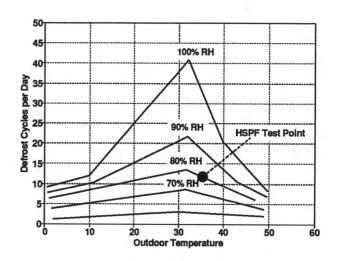

Figure 11-5

Effect of Auxiliary Heater Capacity
During the normal heating cycle, when the outdoor temperature is below the balance point, the supplemental heat will cycle on and off to satisfy the thermostat. In this mode, it does

not matter — from the standpoint of energy consumption — if excessive resistance-coil capacity is energized, because the duration of the operating cycle will be reduced accordingly. However, it matters — from the standpoint of energy consumption — if an excessive amount of resistance heat is energized during the defrost cycle, because this heat will be delivered to the structure at a COP of 1.0; which means that an equivalent amount of heat will not be added by the more efficient device (the heat pump COP is typically greater than 1.0 for most operating conditions) after the defrost cycle is terminated. ("Excessive defrost cycle heat" refers to the difference between the activated resistance-coil capacity and the capacity required to just neutralize the cooling effect during the defrost cycle.)

11-11 The Load-Hour Equations

Some of the technical literature (standards, handbooks, and design manuals) that addresses the efficiency of residential energy conversion equipment suggests that the annual, purchased-energy load can be estimated by using simple equations that process information about the design load, the number of full-load operating hours, and the published efficiency rating. These equations are reproduced here, along with comments regarding the concept of full-load operating hours and the accuracy of the load-hour equations.

Equation Set
Three AFUE equations are used to evaluate furnace performance; one HSPF equation is associated with air-to-air heat pump equipment, and one SEER equation applies to air-to-air cooling equipment. In these equations, HLH refers to the number of full-load heating hours (annual), QH and QC refer to the equipment sizing loads (heating and cooling loads from a **Manual J** calculation, for example), 0.77 is a "correlation factor" that reconciles the output of the equation with a limited set of field test data, and 8,760 is the number of hours in three out of four years.

AFUE Equations
In the following equations, the AFUE rating is used to generate values for the annual energy loads associated with the components of a gas or oil furnace. In these equations, QB refers to the burner input (BTUH), QP refers to the pilot light input (BTUH), and KW refers to the electrical power delivered to the blower.

$$\text{Burner BTU/Yr} = \frac{0.77 \times HLH \times QH}{AFUE}$$

$$\text{Pilot BTU/Yr} = \left[8760 - \frac{0.77 \times HLH \times QH}{AFUE \times QB} \right] \times QP$$

$$\text{Blower KWH/Yr (intermittent)} = \frac{0.77 \times HLH \times QH \times KW}{AFUE \times QB}$$

HSPF Equation
In the following equation, the HSPF rating is used to generate a value for the annual electrical energy load associated with the heating cycle of an air-to-air heat pump. In this equation, KWH/Yr refers to the electrical energy consumed by the indoor and outdoor refrigeration-cycle machinery and the resistance heating coils.

$$\text{Heat Pump KWH/Yr} = \frac{0.77 \times HLH \times QH}{1000 \times HSPF}$$

SEER Equation
In the following equation, the SEER rating is used to generate a value for the annual electrical energy load associated with cooling-only equipment, or the cooling cycle of an air-to-air heat pump. In this equation, KWH/Yr refers to the electrical energy consumed by the indoor and outdoor refrigeration-cycle machinery.

$$\text{Cooling Kwh/Yr} = \frac{CLH \times QC}{1000 \times SEER}$$

Load-hour Values
In the equations already presented, the concept of full-load heating hours and full-load cooling hours are abstractions that attempt to correlate energy consumption for an entire season with a block of full-load hours (FLR). This concept is demonstrated by the following equation.

$$\text{FLH (Hr/Yr)} = \frac{\text{Seasonal Energy Requirement (BTU/Yr)}}{\text{100 Percent Load (BTUH)}}$$

Notice that this equation serves as a mathematical definition of the load-hour concept; it cannot be used to calculate the load hours associated with a proposed design because a value for the seasonal energy requirement would not be available. However, if degree-day information is available, the HLR and CLH values can be estimated by using the following equations.

$$HLH = \frac{\text{Heating Degree Days (base 65)} \times 24}{\text{(65 - Winter Design Temperature)}}$$

$$CLH = \frac{\text{Cooling Degree Days (base 65)} \times 24}{\text{(Summer Design Temperature - 65)}}$$

In regard to these equations: the degree day value associated with any 24-hour period represents the average indoor-outdoor temperature difference for the same 24-hour period (exactly for sinusoidal temperature variations, and with good accuracy for non-symetrical and non-cycling wave shapes). Therefore, division by the base-65 temperature difference value produces a fraction that represents the equivalent full load hours associated with the day. And, when this fraction is multiplied by the number of hours in a day, it is converted into a full-load-hour value for the day. If these calculations are performed on a daily basis, the resulting string of full-load-hour values can be summed to obtain a seasonal load-hour aggregate. (Mathematically, there is no difference in performing calculations on a daily basis and summing daily load hours, or dividing the seasonal degree-day value by the base-65 temperature difference and multiplying by 24.)

Accuracy of Equations

The accuracy of the load-hour equations, as they apply to a specific home, are subject to question because the weather and load models are crude, and because the equipment model applies to a test chamber scenario. Comments about the weaknesses of these models are listed here.

- The energy equation's accuracy cannot be better than the accuracy of the calculation procedure that is used to generate values for the design loads (QH and QC). And since load calculation procedures (as described in **Manual J**) are designed for sizing equipment, they tend to overestimate the maximum applied load. (This situation is acerbated if the load estimate is based on a worst-case design temperature, and on timid assumptions regarding the thermal efficiency of the structural components, the tightness of the structure, the use of shading devices, and so forth.)

- The use of a generic factor (0.77) to compensate for the free heat provided by solar and internal gains cannot apply to specific structures, accept by coincidence. There also is a question about the use of this factor in the procedure that generates the published efficiency value (it tends to increase the HSPF), and the apparently redundant appearance in the load-hour equation (which reduces the applied load).

- Since the HLR and CLR values are based on dry-bulb temperature data, they do not account for the effects of solar gains, internal loads, and latent loads (which vary over a wide range during a day or an entire season). This means that the load-hour concept is site specific (for heating and cooling), because each home has a unique set of seasonal load patterns.

- A discussion of the problems associated with using seasonal, single-value efficiency descriptors to model the performance of furnaces and air-to-air refrigeration cycle equipment is found in Section 11-9. Seasonal, single-value efficiency ratings are not generated for water-to-air equipment and dual-fuel equipment.

11-12 Descriptors and Dynamic Envelope Models

Some software programs use hourly or partitioned bin data to analyze the dynamic behavior of a specific thermal envelope. This type of modeling is desirable because it returns a site-specific estimate for the seasonal energy that must be delivered to the structure. Unfortunately, these same models may use a published efficiency rating to convert the delivered energy requirement into a purchased energy value. (As explained in Section 11-9, there is no reason to expect an equivalence between the on-site efficiency and the test-stand efficiency of fuel conversion equipment.)

11-13 Seasonal Efficiency not Published

Seasonal efficiency ratings are not published for some types of fuel conversion equipment — water source equipment (well water, or earth-coupled piping loop) or dual-fuel arrangements, for example. In these cases, an hourly simulation or a partitioned bin analysis can be used to calculate seasonal energy requirements.

11-14 Validation of Software Products

Typically, the accuracy of a software program is never validated by comprehensive on-site testing. In this regard, there have been efforts to certify the output of the DOE-2 program (which is being used to check the accuracy of other software products), but this effort only applies to a small percentage of the possibilities that can be generated from a matrix of climate zones, construction features, and mechanical systems. (More work needs to be done on residential comfort system simulations; and more testing is required to evaluate a benchmark model's reaction to a representative range of system features and climatic conditions.)

11-15 Comprehensive Calculation Tool

The ultimate calculation tool would be provided by software that dynamically and accurately models all of the climatic, structural, and mechanical characteristics and features of a specific dwelling. In this regard, a basic inventory of items that affect the annual energy load is provided on the following pages by Attachment 11-1. This summary of "rated features" can be used to evaluate the range of application and level of detail accommodated by a particular software product (assessing the validity and accuracy of the modeling effort is another matter).

Attachment 1 — Rated Features

Foundations

Types	Discretionary Parameters	Other Factors and Comments

Slab-on-grade
Crawl space (open, closed)
Basement
Daylight basement
	Overall U-value or R-value
	Insulation location and orientation
	  * slab edge details
	  * foundation wall details
	  * crawl space ceiling
	  * basement ceiling
	Area or perimeter
	Depth of insulation
	Crawl space vents
	  * none
	  * fixed position vents
	  * operable vents
		R-value of soil path
		Soil temperature swings
		Thermal mass effects
		  * walls
		  * floor
		Conduction path
		  * to unconditioned space
		  * to outdoors
		Temperature estimate
		  * unconditioned spaces

Structural Components

Types	Discretionary Parameters	Other Factors and Comments

Wall (frame, masonry, etc.)
Ceilings under attic
Ceiling on exposed beams
Roof ceiling combo
Ceiling floor combo
Floor
Door
	Framing details
	  * overall U-value or R-values
	  * mass and specific heat
	External shading
	Area
		Loads affected by
		  * incident solar
		  * thermal mass (set-back, set-up)
		  * thermal mass (solar gain buffer)
		  * color
		  * latitude
		  * orientation

Glass

Types	Discretionary Parameters	Other Factors and Comments

Windows
Skylights
Doors
Glass block
	U-value (transmission)
	Number of panes
	Frame performance (U)
	Glass shading coefficient (SC)
	Device shading coefficient
	  * interior (summer and winter)
	  * exterior (summer and winter)
	Elevation angle (sky light)
	Fins and overhangs
	Area
		Cooling loads affected by
		  * incident solar radiation
		  * interior mass (storage transient)
		  * structural mass (solar gain buffer)
		  * color
		  * latitude
		  * sun — altitude and azimuth
		  * orientation

Infiltration

Types	Discretionary Parameters	Other Factors and Comments

Framing crackage
Window and door crackage
Unsealed penetrations (various)
Vent and exhaust openings
Fireplace
Ducts in unconditioned space
	Volume and surface areas
	Quality of construction
	Quality of windows and doors
	Infiltration and/or vapor barrier
	Dampered openings
	Duct location
	Duct materials and sealing
	Use of building cavities for airways
	Fuel and venting
	  * combustible or electric
	Combustion chamber
	  * atmospheric (use indoor air)
	  * isolated from indoor air
		Height of structure
		Exposure to wind
		Door traffic
		AC/HR is misleading descriptor
		  * volume versus leakage area issues
		Synergistic effects
		  * crack leakage
		  * duct leakage
		  * return path restrictions
		  * vents and fireplaces
		  * combustion appliances
		  * exhaust equipment

Attachment 1 — Rated Features (Continued)		
Buffer Zones (Unconditioned Spaces)		
Types	**Discretionary Parameters**	**Other Factors and Comments**

All	Insulation
	Leakage
	Temperature
		Temperature in space not known
		  * envelope transmission
		  * tightness and sealing
		  * effect of solar gains
		  * duct system losses

Ventilation		
Types	**Discretionary Parameters**	**Other Factors and Comments**

None
Mechanical (integral with HVAC)
Mechanical (self-contained)
	CFM
	  * no heat recovery
	  * heat recovery (effectiveness)
		Sensible and latent heat recovery

Internal Loads		
Types	**Discretionary Parameters**	**Other Factors and Comments**
Hard wire lighting Plug lighting Plug appliances and equipment Kitchen equipment (electric or gas) Laundry equipment (electric or gas) Occupants Gas fireplace equipment	Installed wattage Type of lighting fixture Efficiency rating for appliance items	Sensible loads Latent loads Venting affects make-up air load Venting affects infiltration load Load schedules
HVAC Equipment		
Types	**Discretionary Parameters**	**Other Factors and Comments**

Fossil fuel or electric furnace
Fossil fuel or electric boiler
Electric resistance heating fixtures
Air-source heat pump
Water-source heat pump (once through)
Water-source heat pump (earth loop)
Hybrid (dual-fuel) heat pump
Electric cooling only equipment
Ground-water, water-coil cooling
	Installed capacity (range)
	  * heating
	  * sensible cooling
	  * latent cooling
	Design CFM (range)
	GPM (water-source heat pump)
	Efficiency rating
	Controls (see page 11-12)
	Vent dampers (furnaces)
		Dynamic analysis required
		  * loads (heat, sensible, and latent)
		  * equipment performance
		  * distribution system performance
		Part-load performance variables
		  * outdoor conditions (DB/WB)
		  * incident solar and cloud cover
		  * wind velocity
		  * time-of-day and time-of-year
		  * condition of entering air (DB/WB)
		  * entering water temperature
		  * control strategy
		Part-load models
		  * heating capacity (fossil, heat pump)
		  * sensible cooling capacity
		  * latent cooling capacity
		  * dynamic balance points (heat pump)
		  * supplemental heat (heat pump)
		  * input energy required (all)
		  * cycling penalties (fossil, ref-cycle)
		Parasitic load models
		  * defrost cycle (air-source heat pump)
		  * crankcase heat
		  * pilot lights
		Auxiliary equipment performance
		  * fans (indoor and outdoor)
		  * pump (water-source heat pump, boiler)

Attachment 1 — Rated Features (Continued)		
Distribution System		
Types	**Discretionary Parameters**	**Other Factors and Comments**

Ducts
Pipes
	Insulation level (R-value)
	Tightness (ducts)
		Transmission losses
		Leakage losses
		Synergistic effect (leakage)
		  * on HVAC equipment and vents
		  * on envelope leakage
		  * on exhaust and appliances
		  * infiltration

HVAC Controls		
Types	**Discretionary Parameters**	**Other Factors and Comments**

Comfort
Operating
Safety
	Set points (indoor temperature)
	Set-up and set-back schedules
	Dead band
	Supplemental heat (heat pump)
	  * no lockout
	  * outdoor air thermostat and setpoint
	  * ramped recovery
	Defrost controls
	  * timer setting (time-temp)
	  * demand defrost
	Crankcase heater control strategy
	Continuous fan operation
		Safety controls
		  * low and high pressure (refrigerant)
		  * fuel train
		  * motor overload
		Operating controls
		  * fan interlocks
		  * pump interlocks
		  * fuel train
		  * defrost
		  * crankcase heater
		  * second-stage heat (heat pumps)

Potable Hot Water System		
Types	**Discretionary Parameters**	**Other Factors and Comments**

Electric resistance
Electric heat pump
Fossil
Solar
Heat reclaim (refrigerant cycle)
	Efficiency rating
	Tank insulation (R-value)
	Insulation on pipes (R-value)
	Fixture performance
		Solar collector
		  * type
		  * area
		  * tilt
		  * orientation
		Solar storage (tank size)

Passive Solar Designs		
Types	**Discretionary Parameters**	**Other Factors and Comments**

Glazing and thermal mass	Aperture area
	  * orientation
	  * area
	Thermal mass
	  * location
	  * surface area
	Solar distribution fraction
	  * percentage absorbed by mass	

Additional Issues		
Types	**Discretionary Parameters**	**Other Factors and Comments**
Zoned systems Off-peak storage systems Active solar systems Earth-berm homes On-site energy generation	Defined by system type	The number of zoned systems and off-peak storage systems is increasing, the other systems are an insignificant percentage of housing stock.

Appendix 1
Glossary

The heating, air conditioning, and refrigeration industry uses a vocabulary that includes specialized terms. The following glossary contains examples of this terminology, and should provide individuals entering the field with a basic knowledge of the language of the industry. This glossary compliments the index, which provides a second collection of key words.

A

Absolute Humidity: The amount of water in a pound of air, usually expressed in grains-per-pound or pounds-per-pound. (One grain of water equals 1/7,000 pounds of water.)

Absorption: A dehumidification process that routes a flow of air through a material that extracts moisture (water vapor) from the air.

Absorption Cycle: A chemical-change refrigeration cycle powered by a heating device (not normally used for residential comfort conditioning).

Absorptive Film: A thin sheet of transparent material that absorbs selected bands of solar radiation and reduces the solar load associated with the window.

Access Door: A door or panel provided in a structure or equipment cabinet that permits inspection and adjustment of enclosed components and devices.

Air Change: The amount of outdoor air required to completely replace the air in a room or building; not to be confused with recirculated air.

Air Changes per Hour: The number of times per hour the indoor air (contained in a room or building) is replaced with outdoor air (by infiltration or mechanical ventilation).

Air Circulation: The natural or forced movement of air within the conditioned space.

Air Cleaner: A mechanical, electrical, or chemical device (usually a filter) that removes dust, gas, vapor, fumes, smoke, and other impurities from the indoor air.

Air Conditioner: An assembly of machinery and devices that control the temperature, moisture, cleanliness, and distribution of indoor air.

Air Conditioning: A mechanical process of treating air to control its temperature, moisture, cleanliness, and distribution on a seasonal or year-round basis.

Air Diffuser: A device (supply air outlet) that mixes conditioned air with room air.

Air Film: The layer of air that is in intimate contact with an interior or exterior surface.

Air Film Coefficient: A representative thermal resistance value (R-value) used to simulate the conduction-convection process that occurs at the interface of a surface and the surrounding air.

Air Handler: A cabinet that contains the blower (and other devices, such as coils and filters) used to move air through a distribution system.

Air — Recirculated: Air that has been drawn from the conditioned space, processed by the conditioning equipment, and reintroduced to the conditioned space.

Air Turnover Rate: The number of times per hour the air contained in a room or building is drawn from the conditioned space, processed by the conditioning equipment, and reintroduced to the conditioned space.

Air — Ventilation: A flow of air drawn from outdoors and discharged into the conditioned space (whether or not it is processed by the comfort-conditioning equipment).

Ambient Air: The air that surrounds the space occupied by an object (the outdoor air for a house or the outdoor refrigeration equipment; the indoor air for occupants and the thermostat; or the air surrounding a duct run, regardless of location).

Anemometer: An instrument for measuring the velocity of a flow of air.

Atomize: Conversion of a liquid into a mist (for example, the liquid processed by oil burner).

B

Barometer: An instrument used to measure atmospheric pressure.

Blow (see "throw"): The distance an air stream travels from an outlet to a point where the velocity is reduced to a specified velocity (see "terminal velocity").

Blower door: A fan-powered device used to pressurize or depressurize a home, and to measure the leakage rate created by the imposed pressure differential.

Blower: A centrifugal fan.

Boot: A fitting used to connect a duct run to an air distribution device (supply outlet or return fixture).

British Thermal Unit (BTU): The amount of heat required to raise the temperature of one pound of water one degree Fahrenheit (approximately, the amount of heat produced by burning one wooden match).

British Thermal Unit per Hour (BTUH): The rate of heat flow, measured in BTU per hour.

C

Change of State: A transformation from solid to liquid, or liquid to gas; and vice versa.

Charge — Refrigerant: The amount of refrigerant required to properly fill a complete refrigeration system.

Chimney Effect: The tendency of heated air or combustion gas to rise to the top of a vertical airway (vent, duct, or stairwell).

Coil: A cooling or heating heat transfer device made of tubing and fins (fluid-to-air) or electrical resistance wire.

Comfort Chart: A graphic summary of the various combinations of temperatures, humidities, and air velocities that make a substantial majority of people feel comfortable.

Comfort Zone: The combination of conditions that make most people feel comfortable (see — "comfort chart").

Compressor: A mechanical device that pumps the refrigerant through the components of a refrigeration system.

Condensate: The liquid formed when a vapor condenses.

Condensation: The liquid produced when a vapor is cooled to a temperature below its dewpoint (heat is extracted during this process).

Condensation Point: The temperature at which the removal of an additional increment of heat initiates the condensation process.

Condenser: A heat transfer device used to convert a vaporized refrigerant into liquid refrigerant.

Condensing Unit: A package of equipment consisting of a motor, compressor, condenser, controls, and associated hardware.

Conditioned Space: The enclosed volume of a room or home served by a comfort conditioning system.

Conduction: The flow of energy, in the form of heat, through a solid material.

Conduit: A tube or pipe used for conveying liquid or gas; or a tube or pipe through which wires or other pipes are routed (for protection against damage).

Convection: The transfer of heat from a fluid or gas to a solid material (or vice versa).

Control: A manual or automatic device for regulating the operation of a mechanical system or device.

Crawl Space: The space between the ground and the floor of a home.

D

Damper: A manually operated (hand damper) or automatic throttling device used for adjusting the flow through a duct run, equipment cabinet, supply outlet, or return inlet.

Diffuser: A supply air outlet designed to provide an aggressive mixing action.

Degree Days: The summation of the average daily temperature differences (usually referenced to 65°F) associated with each day of the heating season (or the cooling season).

Degree Hours: Degree days multiplied by 24.

Daily Temperature Range: The average difference between the daily high temperature and the daily low temperature for a specific location.

Defrost Cycle: The mode of operation associated with removing the ice or frost that — depending on the outdoor temperature and the amount of moisture in the outdoor air — accumulates on the outdoor coil of an air-source heat pump operating in the heating mode.

Dehumidification: The use of mechanical cooling equipment or chemical-process equipment to extract water vapor from a flow of air.

Dew Point: The temperature at which the water vapor contained in the air turns to liquid; the temperature at which the air becomes completely saturated with water vapor (100% relative humidity).

Draft: A noticeable movement of air caused by mechanical or natural forces that is normally considered objectionable; or the flow of gasses through a vent or chimney.

Dry-Bulb Temperature: The air temperature, measured by an ordinary thermometer when there is no solar heating or evaporative cooling effect.

Duct System: A network of tubular or rectangular conduits and connectors (elbows, tees, branch fittings, and boot fittings) used to move air from one point to another.

E

Elbow: A sort, bent conduit.

Enthalpy: The amount of sensible and latent heat contained in a pound of air at a specified set of conditions (usually expressed as BTU-per-pound).

Entrainment: The induction of room air into the flow of high velocity air discharged from a supply outlet.

Equivalent Temperature Difference (ETD): A surrogate temperature differential used to simulate the effects of solar heating and thermal storage when calculating the heat gain through a wall, partition, roof, or door.

Evaporation: The conversion of a liquid to a vapor (heat is absorbed and stored during this process).

Evaporator: The refrigerant-system heat exchanger used to vaporize liquid refrigerant.

Exfiltration: The air flow escaping from an enclosed space through a hole, seam, or crack in the surface of the structural envelope.

Exhaust Opening: An opening through which air is mechanically extracted (by an exhaust fan) from a conditioned space.

Expansion Coil: The refrigerant-system heat exchanger used to vaporize liquid refrigerant (the evaporator).

Expansion Valve: A self powered sensor-valve device used to control the flow of refrigerant entering the evaporator.

F

Fan: A motor driven device for moving air (centrifugal fans are called blowers).

Fan Coil Unit: An air-handling unit (usually containing a blower, one or more coils, and possibly a filter) used to distribute air to a condition space (with or without a duct system).

Filter: Most filters are mechanical devices that can only remove particles and mists from a flow of air. Special filters, such as a charcoal filter, also reduce the concentration of odors and gases.

Filter Drier: A refrigerant piping device, located in the liquid line, that captures particles and absorbs moisture.

Fitting: A connector used to join sections of conduit (runs of duct or pipe), or to join a cabinet to a conduit or a device to a conduit.

Flue: The passage through which gases travel as they move from the combustion chamber to the atmosphere.

Free area: The total open area associated with the face of a grille or register.

G

Gauge (Gage): An instrument for measuring temperatures, pressures, or liquid levels. Also, an arbitrary scale of measurement for sheet metal thicknesses and wire diameters.

Grains: A unit of measurement that refers to the weight of an object. (There are about 7,000 grains in a pound.)

Grille: A supply- or return-air fixture that is not equipped with a damper.

H

Heat: The thermal energy stored in an object or substance. (If an object or substance were devoid of heat, it's temperature would be approximately 460°F below zero (absolute zero).

Heat-Absorbing Glass: Glass that absorbs the energy associated with selected bands of light waves. (It is designed to reduce the amount of summer solar heat gain and to retain heat in the winter.)

Heat Exchanger: A device in which heat is transferred from one fluid to another, such as a condenser, evaporator, or the heat transfer assemblies installed in hot air furnaces and boilers.

Heat Gain: The total heat flow into a space from all sources (people, lights, machines, sunshine, conduction, infiltration, ventilation, and so forth).

Heat Loss: The total heat flow out of the thermal envelope by conduction, infiltration, and ventilation.

Heat Transfer: The flow of energy from a warmer surface, object, or body to the cooler surface, object, or body (by conduction, convection, and radiation).

Heat Transfer Multiplier (HTM): (In **Manual J**) — The amount of heat that flows through 1 square foot of a structural surface when the surface is subjected to a specified temperature difference. (To determine heating HTM for glass, doors, walls, roofs, ceilings, and floors, multiply the transmission coefficient (U-value) by the winter design temperature difference. To determine cooling HTM for glass, doors, walls, roofs, ceilings, and floors, multiply the transmission coefficient (U-value) by the equivalent temperature difference values provided in **Manual J**.)

Heat Pump: A reversible-cycle refrigeration system used to provide heating and cooling.

Heat Transmission: The time-based flow of heat associated with conduction, convection, and radiation.

Heating System, Warm Air: A central heating system that uses duct runs, supply air outlets, and return grilles to distribute heat throughout the home.

Heating System, Perimeter: Perimeter warm-air systems are characterized by supply outlets located in the floor along outside walls, particularly under windows or other surfaces having a high rate of heat loss.

High-Side: The part of a refrigeration system between the compressor and the condenser (the portion of the system that has the highest temperatures and pressures).

Humidifier: A device that adds moisture to warm air being introduced into a space (or moisture can be added directly to the space).

Humidistat: A control device designed to regulate the output of a humidification device.

I

Infiltration: Uncontrolled outdoor air leakage into a conditioned space through cracks and openings in the thermal envelope. (Infiltration is caused by a collection of pressure drivers, including wind, stack effect, vents, chimneys, exhaust fans, and duct leaks.)

Induction: The entrainment of room air into the air stream projected from a supply-air outlet.

Insulation: A material that has a characteristically high resistance to heat flow, used to reduce heat loss and heat gain.

K

K-Value: The heat flow through 1 square foot of a material that is 1 inch thick, when exposed to a 1°F temperature difference for 1 hour.

L

Liquid Line: In a refrigeration circuit; the pipe that connects the condenser or receiver to the pressure-reducing device, located just upstream from the evaporator.

Latent Heat: The heat associated with a phase change, for example from gas-to-liquid, liquid-to-solid, or vice versa. (The temperature of the material remains constant during this process.) In **Manual J**, the time-based heat flow associated with changing (removing or adding) the amount of moisture in a flow of processed air.

Louver: A matrix of vanes used to force a directional change in a flow of air (usually provided with air-intake fixtures to block entry of rain and snow). Louvers also are used on indoor air-transfer-fixtures for privacy and to eliminate light penetration.

Low Emittance (Low-E) Glass: A chemical coating applied to glazing that enhances the thermal resistance (R-value) of the glass without significantly decreasing visibility. (Low-E glass is used to reduce the heating load associated with glass, but there is a minor benefit during the cooling season.)

Low-Side: The part of a refrigeration system between the evaporator and the compressor (the portion of the system that has the lowest temperatures and pressures).

M

Make-Up Air Unit: Equipment that processes outdoor air and introduces it into a space (to replace air exhausted from the space).

Manometer: An instrument that uses liquid level difference to measure pressure.

Multi-Zone System: A comfort-conditioning system designed to serve two or more areas having different heating, cooling, humidification, ventilation, and filtration requirements.

Muffler: A device used to attenuate propagated noise (inserted in duct run or in a refrigeration piping run).

O

Oil Separator: A device for separating oil or oil vapor from a flow of refrigerant.

P

Permeance: A term that refers to the quality of a vapor retarder material (the ability to resist vapor migration caused by differential in vapor pressure).

Perm: A unit of measurement used to quantify a material's ability to retard moisture migration. (One perm corresponds to a migration of 1 grain of water vapor, through 1 square foot of material, in 1 hour, while the material is subjected to a pressure differential of 1 inch of mercury.)

Phase Change: A term that refers to a transformation of a material (solid-to-liquid, liquid-to-gas, or vice versa), also called a "change-of-state."

Plenum Chamber: A box-like compartment used to connect duct runs to a heating or cooling unit, or to interconnect a system of duct runs. Any enlarged section within the air-side of an equipment cabinet.

Pneumatic: A mechanical device operated by air pressure.

Power Roof Ventilator: A motor-driven exhaust fan mounted above the roof on a roof curb.

Pressure, Absolute: That pressure measured with a gauge (PISG), plus the atmospheric pressure (14.7 PSI at sea level and standard conditions). For example, a gauge pressure of 68.5 PISG represents an absolute pressure of 83.2 PSIA (68.5 + 14.7 = 83.2).

Pressure, Atmospheric: Pressure resulting from the weight of the atmosphere which is 14.696 pounds per square inch of surface; or, that pressure in the outdoors or within a space that is present as a result of the forces of nature.

Pressure Drop: That pressure lost between any two points of a piping or duct system due to friction or leakage.

Pressure, Static: The force per unit area acting against the walls of a container such as a plenum or an air duct (the bursting pressure).

Pressure, Total: A sum of the static and velocity pressures.

Pressure, Velocity: The pressure created when a flow of air is stopped by a flat, perpendicular surface.

Psychrometer: A thermometer-like instrument used to simultaneously measure wet-bulb and dry-bulb temperatures.

Psychrometric Chart:

Psychrometrics: The study of the relationship of the properties of air (dry-bulb temperature, wet-bulb temperature, enthalpy, specific volume, relative humidity, and absolute humidity).

R

Radiant Barrier: A surfacing material (usually bright foil) used to reduce the amount of heat transferred by the radiation process.

Radiation: Transmission of energy by electromagnetic waves, for example, heat transmitted by the sun to a roof.

Radiator: A mechanical device that transmits heat by the radiation process.

Recirculated Air: The air extracted from the conditioned space, processed by the comfort-conditioning equipment, and returned to the conditioned space.

Refrigerant: The fluid in refrigeration-cycle equipment.

Reflective Films: A thin film applied to the inside surface of glass that reflects impinging sunlight to the outdoors.

Reflective Glass: A type of glass that reflects impinging sunlight to the outdoors.

Reflective Insulation: Insulation material coated with a foil that reduces radiant heat transfer.

Refrigeration Cycle: The collection of processes associated with converting liquid refrigerant to a vapor (by using a metering device and expansion coil), a vapor to hot gas (the function of the compressor), and hot gas to a liquid (in the condenser).

Register: A grille accessorized with a damper.

Relative Humidity: The degree of saturation of moist air, expressed as a percentage. For example, a 30 percent relative humidity means that the air is 30 percent saturated; a 100 percent relative humidity means that the air cannot hold any more moisture. (The maximum moisture-holding ability of air depends on its temperature.)

Return Air: Air that is extracted from the conditioned space and routed to the conditioning equipment.

R-Value: A measurement of the resistance to the heat flow through a material, when the material is subjected to a temperature difference that does not change with time.

S

Saturation Point: A term used when air contains all the moisture that it can possibly hold (100 percent relative humidity) at a given temperature.

Sensible Heat: The heat associated with the temperature change of a substance (without a phase change).

Sight Glass: A short piece of glass tubing used to assess the condition of the refrigerant in the liquid line, and the liquid level in tanks, reservoirs, and receivers.

Sling Psychrometer: An instrument that contains a set of dry- and wet-bulb thermometers (used to determine relative humidity or the amount of moisture in the air).

Solar Effect, Solar Heating, or Solar Gain: The sun-generated heat transferred to a room or space through the surfaces (glass, walls, roof) exposed to the sun's rays.

Specific Gravity: The specific weight of a liquid, compared to the specific weight of water; or the specific weight of a gas, compared to the specific weight of air.

Specific Weight: The weight of a substance per unit volume, for example, "pounds per cubic foot."

Specific Volume: The volume of a substance per unit of weight, for example, "cubic feet per pound."

State (of Matter): One of the three forms in which a substance can exist (solid, liquid, or gas).

Steam: Water that has been transformed into a vapor by the addition of heat, or by a reduction in pressure, or a combination of the two effects.

Strainer: A device for separating foreign matter from a flow of liquid or gas. (Sometimes a strainer is installed in the liquid line of a refrigerant system.)

T

Temperature: A measure of the molecular energy associated with a substance.

Temperature Difference: The difference in temperature between two points; for example, the indoor-outdoor temperature difference.

Temperature Drop: The temperature change between two points in a system, for example, between the furnace discharge opening and the supply air outlet.

Thermal Envelope: The structural surfaces that form the interface between the conditioned space and the outdoors, or a conditioned space and an unconditioned space.

Thermal Resistance: The resistance to heat flow.

Thermal Storage: The ability of a material to absorb heat.

Thermography, Infrared. An instrument that evaluates a structure's heat loss by a photographic process.

Thermometer: An instrument for measuring the temperature (dry- and wet-bulb) of gases, fluids, and other substances.

Thermostat: An instrument which sense a change in the temperature at a point and uses this information to control the operation of a mechanical device.

Throw: The horizontal or vertical distance that a jet of air travels after leaving an air supply outlet.

Terminal Velocity. The velocity of a jet of air at the end of its throw.

Ton of Refrigeration: The heat required to melt one ton of ice in a period of 24 hours (12,000 BTU-per-hour).

Total Heat: Total heat amount of energy contained in a substance; the sum of the sensible heat and latent heat (see "enthalpy").

Tracer Gas: A harmless gas used to measure the air change rate associated with the thermal envelope of a home.

Trunk Line: One of the primary conduits in a duct system: the conduit used to feed branch lines (supply trunk) or the conduit used to collect the flow leaving branch runs (return trunk).

U

Unit: An assembly of machinery, controls, and cabinetry used for conditioning and processing air.

Unit Heater: A heating device that includes a heating element and a fan.

U-Value: A heat transfer performance index that combines the thermal resistance across a panel with the convective-resistance effects associated with the panel's exposed surfaces (inside and outside). U-values are expressed in BTUH-per-SqFt-per°F.

V

Valve, Charging: A valve used to add refrigerant to a refrigeration system, or to add oil to the compressor crankcase.

Valve, Emergency Relief: An automatic discharge valve that protects against an over-pressure or over-temperature condition.

Valve, Solenoid: A open-close valve operated by an electric coil.

Vapor: A dispersed collection of liquid molecules that behaves like a gas.

Vapor Barrier (See "vapor retarder"): A moisture-retarding material used to control the amount of moisture that migrates from a humid space to a drier space.

Vapor Pressure: The portion of the measured pressure that is attributed to the action of the vapor molecules. (The vapor pressure increases as the concentration of molecules increases.)

Vapor Pressure Differential: The vapor pressure difference across a membrane that separates different concentrations of a molecular substance.

Vapor Retarder: The correct terminology for "vapor barrier." A low-porosity membrane attached to the inside surface of a structural panel to prevent moisture from migrating through the panel.

Ventilation: The process of replacing indoor air with outdoor air (mechanically, by using a ventilation-only system, or by ducting outdoor air to the return-side of an air distribution system; or naturally, through windows and vent openings).

W

Wet-Bulb Depression: The difference between a simultaneous reading of dry-bulb and wet-bulb temperatures.

Wet-Bulb Temperature: The temperature of moving air measured by an ordinary thermometer wrapped with wet gauze. (The evaporation effect causes a depression in the reading that is related to the amount of moisture in the air.)

Appendix 2
Replacement and Retrofit

The design and installation work associated with updating and improving an existing comfort system creates problems not encountered in new construction. For example, design options related to system concept and equipment type are typically limited by the existing equipment arrangement and distribution system layout. In addition, installation work is more difficult when there is limited access to spaces that normally would be used to set equipment, or to route ducts, pipes and wire. Furthermore, a considerable amount of time and effort must be devoted to inspecting the existing system, documenting these observations, and developing a plan of action regarding salvage, modification, replacement, and new construction.

A2-1 Classification of Retrofit Work

Upgrading an existing comfort system will normally fall into one of the categories discussed in this section. As indicated in the balance of this appendix, none of this work is routine, and some of it can be exceptionally challenging.

Equipment Replacement
When wornout equipment must be replaced, the contractor should not be lured into a feeling of security simply because a forced air system is already in place. For example, if the performance of the previous system was unsatisfactory because of a design flaw or installation error, a simple exchange will not resolve the problem. And in some cases, the exchange will create a new problem, or acerbate an existing, borderline problem. (A classical example is provided by the humidity-control problems that are created when older, oversized, low-efficiency cooling units are replaced with equally oversized, high efficiency packages.)

System Upgrade
Homeowners typically upgrade an existing forced air system to improve comfort and lower operating cost. This work might involve adding a function not currently provided (cooling and dehumidification, humidification, filtration, or ventilation), modifying the system concept (adding zone control or replacing air-source equipment with water-source equipment, for example), or simply exchanging an inefficient equipment package with a state-of-the-art product.

New Forced Air System
The reasons for adding a forced air system to a home served by another type of system are based in a desire to improve comfort, and to reduce operating cost. A classical example is provided by a home that has no cooling, and electric base-board heat. (A heat pump would be more efficient, and it adds summer cooling, filtration, and ventilation capability.)

A2-2 Survey

While there is little room for error when designing and installing a system for a new home, there is even less room for error when making major modifications to an existing residence. Therefore, a methodical approach is required to insure success. In this regard, each retrofit project should begin with a detailed survey that can be used as a basis for making design decisions, generating cost estimates, and performing calculation procedures. Of course, the amount of detail will depend on the nature of the retrofit project, which may involve heating, cooling, or adding accessories.

Heating System Survey
The heating system survey should include a scale drawing of the floor plan, complete with sketches and notes regarding construction details and equipment arrangements. This documentation should summarize information (such as location, type, size, capacity, and performance index values) about windows, doors, insulation, structural tightness, exhaust fans, appliances, supply outlets, return grilles, duct runs, piping runs, and the space available for equipment and distribution systems. Also:

- Determine why the existing equipment is no longer functional, and the cause of damage that cannot be classified as normal wear.

- Record comments made by the owner about faults in the old system.

- Make notes regarding structural upgrades that might be justified by an acceptable return-on-investment or a desirable improvement in comfort.

- If there is an existing duct system, inspect the entire system and note the sections or locations that must be insulated, sealed, or replaced.

- Do not oversize the new equipment or the replacement equipment — base the output capacity on a detailed load calculation that simulates the conditions that will exist after the project is finished.

- Be alert for the advantages of using a different type of furnace or equipment arrangement, including the use of an outdoor (rooftop or on-grade) unit.

- If there is a need for duct system modifications or air distribution hardware changes, make sure that the structural systems, and the other building systems will accommodate the work.

- Investigate the availability and cost of fuels and determine if the electrical power supply will be adequate.

- Anticipate and resolve the system problems associated with adding functions (cooling, improved filtration, and power humidification) to an existing system.

- Determine if the vent or chimney is the correct type and size for the fuel being used, and make sure that the vent pipe connector conforms with codes and standards.

- Determine if there is an adequate source of combustion air, or develop a plan for obtaining the combustion air.

- Determine if fireproofing is necessary to comply with codes and regulations.

- Make sure that the new equipment can be carried in through existing doorways (or windows), and stairways.

- If noise has been objectionable with the old system, determine how to eliminate the problem.

- If the old unit is a gravity system, or some other outdated system, consider replacing the duct system with a new distribution system.

- To avoid sound problems when replacing a gravity furnace, locate the new forced air furnace away from a large return duct or grille.

- When setting the new equipment, select a location that facilitates service work.

- All wiring should comply with codes and regulations.

- Search for deterioration (corrosion and rusting) that indicate the existing furnace, flue, or vent were subjected to a hostile environment (chlorinated molecules in the air or insufficient combustion air, for example).

- Search for deterioration (corrosion and rusting) that indicate a venting problem (that caused excessive condensation in the furnace and vent system).

- Search for deterioration (corrosion and rusting) that indicate contaminated fuel (sulfur, for example).

- Make sure that crawl spaces are insulated and sealed (or vented, depending on the situation), and that a concrete slab or a properly installed vapor barrier covers the floor.

Cooling System Survey
The cooling system survey is similar to the heating system survey, especially in regard to the floor plan drawings, supplemental sketches, and notes required to comprehensively document existing conditions. Some items and concepts that pertinent to cooling are listed here.

- Determine why the existing equipment is no longer functional, and the cause of damage that cannot be classified as normal wear.

- Record comments made by the owner about faults in the old system.

- Make notes regarding structural upgrades that might be justified by an acceptable return-on-investment.

- If there is an existing duct system, inspect the entire system, and note the sections or locations that must be insulated, sealed, or replaced.

- Do not oversize the replacement equipment or the new equipment — base the installed capacity on a detailed load calculation that simulates the operating conditions that characterize the finished project.

- A complete change-out (new indoor and outdoor units) is preferable to a condensing unit replacement because refrigeration-cycle equipment will not provide the desired performance unless the components are expertly matched to each other.

- If there is a need for duct system modifications, or air distribution hardware changes, make sure that the structural systems, and the other building systems will accommodate the work.

- If the home features a hot-water heating system or steam heat, a complete air distribution system will have to be installed in the available space (attic, basement, or crawl space).

- If a cooling-only distribution system is installed in the attic, be sure to provide means to close off the ceiling diffusers and the returns during the heating season. Failure to isolate this system will cause objectionable drafts during cold weather. In extreme cases, moisture could drip from supply outlets and returns, and water could accumulate in the duct system and air-handler.

- Whenever a fan-coil unit is located in an attic, it should be installed above a water-tight pan that is equipped with a drain-pipe.

- Ductless, split-coil equipment may be required when it is economically or technically impractical to add forced-air cooling to an existing heating system.

- If cooling is added to a forced air heating system, the condition of the existing furnace must be compatible with the new cooling equipment.

- If a cooling coil is added to a furnace, the air handling capabilities of the existing blower must be compatible with the additional air-side resistance.

- If cooling is added to a forced air heating system, the air handling capabilities of every component of the duct system must be compatible with the new cooling equipment.

- If the existing system has a counterflow furnace with ducts below a slab, the pipe sizes must be compatible with the air flow requirements of the new cooling system.

- If the blower seal in an existing furnace is not tight at the division panel between the heating and return sections, the problem should be corrected.

- Outdoor refrigeration-cycle equipment makes noise and expels heat, so it must be set in a position that will not disturb the homeowner and neighbors, or damage shrubbery.

- Condensers reject heat to the surrounding environment, so do not locate this equipment where it is unusually hot or where air cannot freely circulate through the equipment.

- Anticipate and resolve the piping and architectural problems associated with condensate disposal.

- Certify that the electrical service is compatible with the increased load.

- Make sure that crawl spaces are properly insulated, sealed (or vented, depending on the situation), and that a concrete slab or a properly installed vapor barrier covers the floor.

- Make sure that the attic is properly vented.

- Examine the existing thermostat and control wiring to determine if it will function as a combination heating and cooling thermostat, and replace if necessary.

- Make sure that all registers, diffusers, and grilles are of adequate size to deliver the required air flow, with a discharge pattern that is appropriate for the intended duty.

- Make sure that the return system is adequate for the intended duty.

- Consider how furnishings and draperies will affect the performance of the supply air outlets.

- It may not be necessary to use existing basement outlets during the cooling season; when this is the case, design accordingly.

- Make sure that there is enough room to add a coil section to an existing furnace, and that there is room for installing refrigerant lines and a drain.

- Fan and limit controls mounted high in the supply plenum, or in the duct, may have to be relocated below the cooling coil.

- Consider installing some type of attenuation if a noise problem is caused by the increased blower speed required for cooling.

- If noise has been objectionable with the old system, determine how to eliminate the problem.

- If equipment is placed on the roof, consider the rigging required, the accessibility for service, the type of curb or support required, the load-carrying qualities of the roof structure, and how refrigerant and power lines will be installed.

- If equipment is placed on the ground, use a concrete pad.

- Make sure that the new equipment can be carried in through existing doorways (or windows) and stairways.

- Refrigerant lines should penetrate the structure at points that are above grade, and they should be properly isolated and insulated (suction line).

- If there is substantial difference between the elevation of the condensing unit and the evaporator, consult manufacturer's recommendations pertaining the refrigerant piping system.

- All wiring should comply with codes and regulations.

Accessory Item Survey
Accessory air filters and humidifiers are frequently added to an existing comfort system. Use the following check-list to evaluate the factors that affect this work.

- Before adding an electronic air filter (or any other type of accessory filter) be sure that the air-handling blower will have the power to overcome the additional resistance.

- If the accessory filter has a water-wash feature, consider the problems associated with providing a water supply, and determine how to dispose of the waste water.

- Consider the size and location of the air-handling cabinet and the spatial requirements of the accessory air filter, including the clearances required for cleaning and service.

- Never install an electronic air filter on the supply side of the air distribution system.

- Make sure that accessory filter is installed so the direction of air flow coincides with the intent of the design.

- Electronic air cleaners, and other types of accessory filters, should be sized and installed in accordance with the manufacturer's instructions.

- If winter humidification is desired, collect the information that will be used to estimate the humidification load.

- Before selecting humidification equipment, refer to the manufacturer's engineering information.

- Certain type of humidifiers require a controlling humidistat to prevent runaway humidification.

- Determine if the water supply is compatible with the humidification equipment, as far as function and maintenance is concerned.

- Install the humidifier in a plenum or section of trunk duct compatible with a source of electric power, the point of water supply, and a drain.

- Exercise caution when installing by-pass humidifiers; moist air may cause electronic air filters to arc excessively. Locate the bypass connection on the discharge side of the air cleaner.

- If a bypass humidifier is added to a system that provides cooling, put a manual damper in the bypass duct and close the damper during the cooling season.

- Power humidifiers must be shut down and drained during the cooing season.

A2-3 Final Design

After the survey information has been collected, the system concept can be finalized and the design work can begin. In this regard, the design procedure is identical to the procedure used for new construction, as summarized below.

- Evaluate the zoning requirements.

- Evaluate the tightness of the structure and make provision for ventilation air or combustion air, if required.

- Use **Manual J** to calculate the loads (heating, sensible cooling, latent cooling, and humidification) associated with the existing thermal envelope (no insulation or sealing work, same floor plan) or the improved thermal envelope (tighter construction, added insulation, added floor space), as required by the remodeling plan.

- Produce load estimates for the entire house or for the entire zone, as required.

- Generate a set of load calculations for each room of the house.

- Use the procedures documented in **Manual S** to select and size equipment.

- After the new equipment has been selected, extract the blower performance data from the manufacturer's engineering literature.

- Ignore the survey information regarding existing duct sizes, and use the procedures documented in **Manual D** to determine the size of each existing section of duct and any new duct runs.

- Compare the size of the existing duct runs with the calculated sizes, and mark the sections that are too small (too much flow resistance or high velocity).

- Resolve the differences between the desired duct sizes and the existing duct sizes. (Plan to replace inefficient fittings and undersized duct sections, or find equipment that has a more powerful blower.)

- Plan to add balancing dampers at branching trunk ducts, and to each section of run-out duct. Balancing dampers also may be required at critical points in the return system.

- Evaluate the location and performance of the existing supply air outlets and return grilles, and replace with appropriate hardware (style and size) as required (refer to **Manual T** for selection guidelines).

- If applicable, select terminal hardware for new supply and return runs.

- Make plans to balance the system.

A2-4 Logistics and Pricing

Replacing and installing equipment in existing homes presents a distinctly different set of scheduling and pricing problems. Longer timelines and higher margins are usually associated with this type of work.

- Retrofit work requires a thorough understanding of the existing conditions, which requires inspection time at site.

- Each job is unique, which means that more supervision will be required at the site, and plans may have to be altered when unanticipated conditions are discovered.

- The work may have to be scheduled around the occupants' use of the home, which means that it could be subject to scheduled or unexpected interruptions.

- The work may have to be coordinated with a remodeling or insulation contractor.

- The work may have to be coordinated with the installation of a new power service, a new electrical system, a new piping system, or a new fuel supply system.

- Demolition work is normally required, which means that old equipment, distribution system scrap, and structural debris must be promptly and carefully removed from the premises.

- Building codes may have changed since the original installation, preventing the use of some components in their existing location, and new energy codes may mandate a substantial increase in efficiency.

- With the noted differences between new construction and retrofit work, it should be readily apparent that a higher gross profit margin is justified because of the higher risk and complexity of the retrofit work.

Appendix 3
Provisions for Testing and Balancing the System

If the air distribution system is not designed and installed correctly, it may be difficult or impossible to balance. Even if the system is well designed, it may be hard to balance if test points are not readily accessible to the balancing technician. The difficulties associated with performing air-side balancing work can be minimized by including sufficient access points, by specifying hardware that will expedite the balancing work, and by providing documentation that summarizes the system designer's performance expectations.

A3-1 Design a System That Can Be Balanced

It may be difficult to balance the system if unstable, unpredictable, and inconsistent flow rates are caused by poor entrance and exit conditions at the fan. Inefficient fittings and airway obstructions also can create balancing problems. For example, unstable flow conditions occur when a branch take-off is located in the turbulent wake that develops behind a poorly designed fitting, or immediately downstream from an air-side device.

- Elbows, tees, and branch take-off fittings should not be located close to the cabinet discharge opening. For the range of blower sizes used with residential duct systems, there should be at least three feet of straight duct between the discharge opening collar and the first fitting.

- A properly designed coil section can be installed at the discharge opening of a furnace cabinet. However, there should be at least three feet of straight duct between the coil-section discharge collar and the first duct fitting.

- The flow-through heating coils and cooling coils should be uniform across the face of the coil. A gradual transition from the cabinet opening to the coil-section opening is required when the coil size is not compatible with the size of a cabinet discharge collar.

- Return-branch fittings should not be located near the return-side of the blower. Make sure that branch duct reentry fittings are at least four feet from the return opening collar.

- Tees, elbows, and branch takeoffs should be designed to help the air move through the turn (fittings that feature turning vanes, radius corners, or a 45 degree entry are very desirable).

- Tees should be designed so that the geometry of the turning vanes or radius elbows mechanically divide the flow into two streams that have the desired percentage of the upstream flow.

- Turning vanes should be installed so that the leading edge points directly upstream and the trailing edge points directly downstream.

- Supply air outlets should never be installed flush on the duct wall. Collars or necks are required, and equalizing grids are recommended. Extractors can be used, providing the next branch take-off fitting is at least four equivalent duct diameters downstream from the turbulent wake.

- Refer to the duct construction standards published by the Sheet Metal and Air Conditioning Contractors National Association (SMACNA) for elbow and tee design details, and for grille and register installation details.

- Flexible duct should be cut to length and installed so that the duct run is straight. Sags, coils, and crimps are not acceptable.

A3-2 Balancing Dampers Required

Balancing dampers must be installed at locations that allow the balancing technician to have complete control over every aspect of air-side performance. Trunk-duct dampers are useful for balancing systems that feature multiple trunk runs, and branch-duct dampers should be used for adjusting the air flow through supply outlets and returns. (Because of noise, register dampers are not recommended for this task.)

- On the supply-side, a straight duct that is at least three feet long should be installed between the discharge collar and a trunk balancing damper. On the return side, a trunk balancing damper should not be installed within four feet of a return opening collar.

- Install opposed blade balancing dampers at all trunk junctions where the air is split into two or more flows. These dampers should be located four to six equivalent trunk diameters downstream from a turbulent wake generated by a fitting or air-side device. (The equivalent diameter of a rectangular duct can be read from a duct sizing slide rule.)

- Install quadrant balancing dampers in every branch take-off. These dampers should be located four to six equivalent branch diameters downstream from a branch take-off fitting.

- Branch take-off fittings should not be located immediately behind a fitting or air-side device that creates a turbulent wake. This means that branch take-off fittings should be installed at a point that is at least four equivalent trunk diameters downstream from the source of a disturbance.

- Because they cause noise, balancing dampers should not be installed near a supply air outlet or return inlet.

- Balancing the return system is just as important as balancing the supply system. Install balancing dampers upstream from all junctions and branch reentries.

- Splitters and extractors can be used to divert the direction of the flow, but they should not be used as balancing devices.

- Do not use registers to balance the system.

A3-3 Provisions for Performing the Balancing Work

The system designer can expedite the balancing work by producing a design that provides access to the air handling equipment and to all critical test points. If the budget allows, pressure taps and sealable access holes can be incorporated into the sheet metal work.

- Provide pitot tube access holes for measuring the blower CFM, and the flows associated with trunks and branch run-outs.

- Provide pressure tap connections for measuring the pressure differential across the blower section or the pressure drop across a coil.

- Never specify a test point, or install a testing device in a section of duct that is subject to distorted or unstable flow patterns. (Access points should be located at least four equivalent diameters from a fan-cabinet collar, a fitting, or the end of the duct.)

- Provide test wells for measuring the temperature difference across a coil or heat exchanger.

- Make sure that there is enough room (around the components of the duct system and equipment cabinets) for the balancing technician to make measurements and take readings.

A3-4 Document the Design

The testing and balancing work will be expedited if the desired performance is fully documented. This information should be collected as the design is executed, and copies should be made available to the balancing technician. Some examples of the type of information that will be useful during the balancing work are listed below.

- Provide information pertaining to the desired flow rate at all supply outlets and returns.

- Provide performance tables that relate blower speeds to flow rates and external pressure. Also provide information about the direction of rotation, voltage, and amperage.

- Provide tables or curves that relate the flow rate to the pressure drop across an air-side device (a filter, coil, or heat exchanger, for example).

Appendix 4
Vibration and Sound Control

Mechanical devices have moving parts that cause vibration and generate noise, but equipment manufacturers have found ways to eliminate, suppress, or minimize these disturbances. However, when these efforts are not 100 percent effective, additional measures are required to guarantee a pleasant environment for the homeowner. It also is necessary to isolate equipment, conduit materials, and structural components.

A4-1 Best Location for Indoor Equipment

Noise generated by machinery is more objectionable in bedrooms, dens, studies, and sitting rooms than it would be in a kitchen or family room; it may go unnoticed in a utility room or shop. Therefore, the equipment should be located as far away as possible from the quiet rooms. In this regard, a basement, crawl space, utility room, or attached garage is preferable to a closet within the living space.

A4-2 Split Cooling Systems — Outdoor Unit

Air-cooled condensers are usually installed outdoors, in close proximity to the structure. When evaluating possible locations, consider the relationship to the quite rooms and the effect on the neighboring structure; then select the least obtrusive location as far as noise and esthetics are concerned. (The location of the outdoor unit is even more critical when the neighboring building is not air conditioned, because equipment noise is easily transmitted through open windows. Care also should be taken to prevent the discharge of hot air toward a neighbor's windows or shrubbery.)

If the equipment is installed on a roof-mounted platform, it should be located away from windows and outdoor sitting areas (porches and patios, for example). This platform also must be of a shape and size that provides a suitable workspace for the service technician.

A4-3 Isolation From Structure

Free-standing furnaces, refrigeration-cycle equipment, and ancillary devices (such as a water pump) should be isolated from the platform that supports the equipment. In this regard, wood floors are resilient, and they have relatively small mass, so adequate isolation is extremely important. (If the equipment is set on the floor of an attic, it should be located at, or near the stiffest part of the supporting structure, which would be above the matrix of wall-studs associated with a bathroom, closet, or hall.) Conversely, if the equipment is installed on a concrete floor (in a basement, crawl space, utility room, or attached garage, for example), vibration isolation is less of a problem.

Equipment also can be suspended from floor joists and roof trusses, providing the supporting structure is of sufficient strength and rigidity. (Use rails or a suspended platform, and try to spread the load over as many structural members as possible). It also is very important to use tuned-spring devices to isolate the equipment, or the equipment platform, from the supporting structure — which may be somewhat bouncy.

A4-4 Equipment Pad

When equipment is set on the ground (outdoors, or in a crawl space), it should be supported by a concrete pad. In addition, an isolation device should be installed between the base of the equipment and the pad.

A4-5 Vibration Isolators

A vibration isolator is selected to accommodate the weight that is concentrated at the point of support. In this regard, the effectiveness of the isolation effort increases as the stiffness of the isolating device is reduced. (A noticeable and equal deflection should be visually apparent at each point of support.)

For residential applications, the most common isolation devices include cork, various types of elastic pads, springs, and spring-damper assemblies. Refer to the manufacturer's literature for information about sizing and installing a particular device.

A4-6 Flexible Connections

Noise and vibration are readily transmitted through rigid conduits. Regardless of the location of the equipment, a flexible connection should be provided at the interface to a pipe run, refrigerant line, condensate line, electrical connection, or duct run.

A4-7 Hangers, Breaks, and Reinforcement

The vibration in a conduit can be transmitted to the structure, and abrasion can damage the conduit material; so it is important to prevent direct contact between a pipe or duct, and a wall, floor, ceiling, foundation, or structural member. This means that hangers, supports, and wall sleeves must be used to isolate distribution systems from the structural components of the building.

Also note that noise can be generated by sheet metal work that does not comply with industry standards. Rumbling, rattling, and oil-canning will not be a problem if the gauge of the wall material is appropriate for the application, and if breaks and reinforcement are used to stiffen wall panels. (Exterior or interior insulation also has a dampening effect on panel vibrations.)

A4-8 Airborne Equipment Noise

The most effective way to control the airborne noise produced by indoor equipment is to install the machinery in a room that has massive structural properties and no leakage paths to the living space (a concrete vault would be ideal). In this regard, a living-space closet equipped with a louvered door can be expected to provide a negligible amount of containment.

If massive structural panels are not an option, the sound that escapes from an equipment room can be reduced by lining the walls and ceilings with sound absorbing material. This strategy will be even more effective if the wall-studs are offset so that alternate studs are in contact with the inside finish of the adjacent rooms. (If a door connects the living space with the equipment room, it should be heavy and tight.)

A4-9 Airborne Noise in Ducts

The sound generated by mechanical equipment can propagate through the supply- and return-side of the duct system. This noise can be adequately attenuated by selective use of sound absorbing materials, by avoiding line-of-sight connections, and by using fittings that do not generate a significant amount of turbulence. (Refer to **Manual D** for more information on this subject.)

- Duct liner and duct board materials have desirable attenuating properties, which can be used to control the amount of airborne noise that reaches the conditioned space. This normally involves installing 6 to 8 feet of acoustic material immediately upstream and downstream from the equipment cabinet. (In some cases, it may be necessary to install sound absorbing material only at the return-side of the system.) In addition, a significant amount of attenuation is provided when sound absorbing material is applied to the inside wall of an airhandler cabinet, plenum, trunk duct, elbow, or tee.

- The use of two or more right-angle turns in a duct run effectively reduces the amount of noise that reaches the conditioned space (especially if the turns feature lined, metal fittings, or fittings fabricated from duct board).

- Aerodynamic fittings are preferred because the turbulence created by inefficient fittings is characterized by broadband noise that can propagate into the conditioned space.

A4-10 Noise Generated at the Point of Distribution

Supply outlets and returns must be properly sized, or noise will be generated at these points. A throttled register damper also will produce objectionable noise. (Refer to **Manual T** for more information on this subject.)

Appendix 5
Pipe Sizing

The family of HVAC piping systems includes refrigerant piping, water piping, gas piping, oil piping, and steam piping. This appendix provides an overview of the work associated with designing various types of piping systems. This work typically includes some or all the following tasks, depending on the type of piping system.

- Calculate flow rates
- Locate and size pipe runs
- Select pipe fittings
- Select and size specialty hardware
- Calculate pressure drops
- Estimate pump performance requirements
- Specify pump RPM, motor power, and impeller size.

Other tasks associated with piping system design include the work related to specifying fabrication methods and selecting materials. In all cases, the system components and the materials must be selected for the intended duty, and they must conform to the applicable code. The design also must allow for expansion caused by temperature changes; and hangers, supports, and vibration isolators must be installed at appropriate points.

Piping systems may be exposed or hidden, depending on the theme of the surrounding (interior or exterior) decor, the available space, and the amount of money budgeted for the project. In addition, there must be no spatial conflicts between a piping system and other piping systems, or other distribution systems (air, electrical, phone, or sound). There also must be no spatial conflicts between a piping system and the structural components of the building.

A5-1 Refrigerant Piping

Since the refrigerant changes phase as it circulates through the system, the refrigerant piping can be thought of as three separate piping systems. The pipe between the compressor and the condenser must be designed to carry a hot, high-pressure gas; the pipe between the condenser and the metering device must be designed to carry a warm, high-pressure liquid and the pipe between the evaporator and the compressor must be designed to carry a cold, low-pressure gas. These three pipe runs are commonly referred to as the hot gas line, the liquid line, and the suction line.

When the refrigerant piping is part of a single-packaged system, there is no design work to do. If a split system is required, the system designer must lay out and size the refrig-

erant lines. Some of the calculations associated with this work relate the parameters that characterize the performance of any type of piping system — flow rates, friction rates, and pressure drops, for example. But, the designer also must be concerned with issues and requirements that pertain only to refrigerant piping systems.

- The piping pressure losses must be small. Large losses will seriously degrade the thermal capacity and efficiency of the refrigeration system.
- The piping design must insure that the oil carried away from the compressor by the refrigerant is returned to the crankcase at an equivalent rate.
- There must be no significant accumulation of oil in any part of the system, other than the compressor crankcase.
- Liquid refrigerant must not enter the compressor.
- Oil should not drain back and accumulate at the head of an idle compressor.
- Refrigerant should not migrate to the compressor crankcase when the system is shut down.
- The refrigerant should not flash to gas while it is in the liquid line.

Refrigeration Line Pressure Drop Data

Tables and charts can be used to estimate the size of refrigerant line pressure drops. Different sources present this data in different formats, but the easiest format to use relates the pressure drop to refrigeration capacity (tonnage), pipe size (diameter in inches), and pipe length (feet of equivalent length). On the next page, Figure A5-1 provides an example of this type of chart. (Three separate charts are required: one for the hot gas line, a second for the liquid line, and a third for suction line. The data also depends on the type of refrigerant. Separate chart-sets have been developed for each type of refrigerant.)

Note that the data presented in Figure A5-1 is based on a 2°F pressure drop. This may seem like a strange way to describe a pressure drop, but it is used because there is a loose relationship between the pressure and temperature of a saturated refrigerant. For example, saturated R-22 has a pressure of 68.55 PSIG at 40°F and 61.25 PSIG at 35°F. In this case, a pressure drop of 5 degrees (in temperature units) corresponds to a pressure drop of 7.3 PSIG (in pressure units).

Also note that the data in Figure A5-1 applies to a specific set of conditions, which are a 2°F pressure drop, a 40°F suction temperature, and a 105°F condensing temperature. This data must be adjusted if the actual design parameters deviate from these values. This is accomplished by applying correction factors, which are included with the chart.

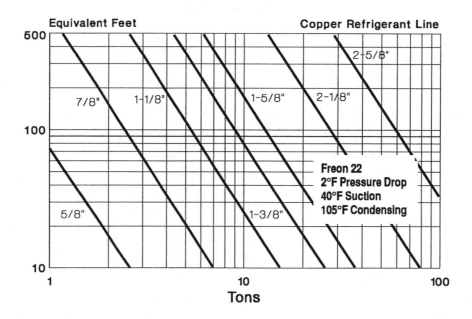

Figure A5-1

Using Refrigeration Line Pressure Drop Data

The pipe sizes can be read directly from the refrigerant line sizing charts. Simply enter the chart at the tonnage value (corrected for suction temperature and condensing temperature, if necessary) and read up to the pipe size and left to the equivalent length values (corrected for pressure drop, if the pressure drop is not 2°F).

The goal is to select a pipe size that provides an equivalent length that is equal to, or greater than, the actual equivalent length of the pipe run. This requires a trial and error solution. Begin by selecting a pipe size that provides an equivalent length that is equal to, or a little larger than, 1.5 times the measured length. Use this tentative pipe size to lookup the fitting lengths, and to calculate the total equivalent length of the run. If this calculated length is too long or too short for the tentative pipe size, select the next larger or smaller size, and repeat the calculation. (Piping manuals provide tables that list the equivalent lengths of copper and steel pipe fittings, valves, and other piping hardware.)

Suction Line Sizing

Suction lines are commonly sized for a 2 degree pressure drop. This is a simple procedure, providing there is no riser. However, if there is a riser (because the compressor is above the evaporator, or the line loops up before dropping to the compressor), the designer must make sure that the oil will circulate with the refrigerant when the compressor is operating. Information about suction riser design is provided in this section.

Hot Gas Line Sizing

Hot gas lines are commonly sized for a 2 degree pressure drop. Again, this is a simple procedure, providing there is no riser. If there is a riser (because the condenser is above the compressor, or the line loops up before dropping to the condenser), the designer must make sure that the oil will circulate with the refrigerant when the compressor is operating, and that oil will not drain back to the compressor when it is shut down. Information about hot-gas riser design is provided in this section.

Suction and Hot Gas Risers

If a riser is required, the hot gas must move up the riser with enough velocity to pull the oil up, along the wall of the pipe. The required riser size can be determined by referring to refrigerant piping tables that correlate riser size with tonnage. When using these tables, the tonnage "look-up value" is equal to the minimum operating tonnage, which will be established by the capacity control hardware.

Once the riser is sized for the minimum load condition, it is necessary to recheck the pressure drop across the entire run, at the full load condition. If the pressure drop associated with the entire suction or hot gas line exceeds 2°F, the horizontal pipe sizes can be increased to compensate for the pressure drop associated with the riser.

Also note that a hot gas line should be looped below the compressor before it turns up to the riser. This loop (see Figure

A5-2) provides a reservoir that can hold the oil that drains down the riser when the compressor is off. This arrangement prevents the oil from reaching the compressor's head.

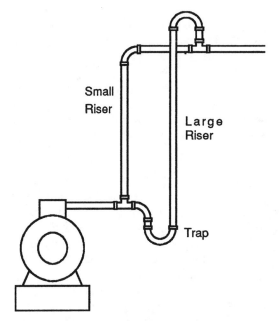

Double Hot-Gas Riser

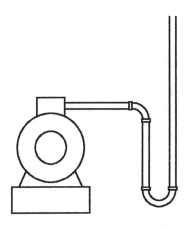

Figure A5-2

Double Risers
In some cases (particularly if the compressor unloads to a very low capacity), it may be is necessary to install a double suction riser or a double hot gas riser. (When the compressor operates at low capacity, oil fills the trap at the bottom of the large riser and plugs the riser. When this occurs, the hot gas that is forced through the small riser moves fast enough to drag the oil up, along the wall of the pipe. In this mode of operation, the pressure drop is not excessive because the gas flow rate is low. Alternatively, when the compressor is loaded to the upper range of its capacity, the oil will be blown out of the trap, and gas will flow up both risers. In this mode, there is more flow available to drag the oil along the walls of the large and small risers, but since there is more pipe to carry the flow, the pressure drop is not increased. Refrigerant piping tables are used to size the pipes that provide the double riser arrangement.)

Liquid Line Sizing
The liquid line also can be sized for a 2°F pressure drop. But, this 2°F pressure drop accounts only for the resistance of the pipe and fittings. Other pressure drops may be associated with refrigerant line accessories, such as valves, filters, and driers. These pressure drops must be added to the drop associated with the pipe run. And, if there is a liquid line riser, there will be a hydraulic pressure associated with the column of liquid that fills the riser. (If the direction of flow is up, this hydraulic pressure must be added to the other pressure drops, but if the flow is down, it is subtracted from the other pressure drops.)

Refrigerant line accessory manufacturer's catalogs provide information about the pressure drops associated with their products. Refrigerant piping manuals provide information about the head pressure associated with a column of refrigerant. Since this information is normally provided in PSIG units,

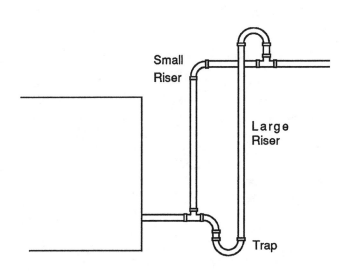

Double Suction Riser

the liquid line calculations also are made in PSIG units. When required, refrigerant properties tables can be used to correlate a saturated temperature with a saturated pressure.

Once the liquid line pressure drop is calculated, the designer can use this information to determine how much sub-cooling will be required to prevent flash gas from forming in the liquid line. This is accomplished by subtracting the liquid line pressure drop from the saturated condensing pressure. (The output of this calculation will be equal to the pressure at the entrance to the expansion valve.) A refrigerant properties table can be used to convert this pressure to a saturated temperature. The smallest amount of sub-cooling that will prevent flashing is equal to the difference between the saturated condensing temperature and the temperature at the entrance to the expansion valve. In practice, a few extra degrees of sub-cooling are used to insure that no flashing will occur.

Oil is readily returned through liquid lines because oil readily mixes with a liquid refrigerant. Since the oil is extensively dispersed throughout the mixture, it cannot accumulate in the low spots of the pipe run.

Detail Piping

If a split system is used, the detail piping is usually part of the equipment package. If detail piping is required, get specific piping instructions from the manufacturer, especially if more than one compressor and/or evaporator is piped into a single refrigeration circuit. Incorrect detail piping can cause serious problems, including incorrect and unbalanced operating pressures and inadequate oil return.

Codes and Regulations

Sizing procedures and stipulations about methods and materials are extensively codified. Make sure that the design conforms to building codes, fire codes, EPA regulations, and insurance underwriter requirements.

A5-2 Water-Piping Systems

When a water-piping system is designed, the pipe runs are sized first and then the system pressure drop is estimated. This information is used to select a pump that will provide the desired flow and pressure. After this work is completed, the expansion tank and other accessories can be selected. A brief overview of this design procedure is provided below.

Flow Rate Values

If the terminal flow rates are noted on a drawing of the pipe system, the GPM flowing through any section of the system is determined by totaling the GPM values associated with the mains and branches downstream from the section of interest. These calculations are made by starting at the ends of the pipe system and working back to the pump. The accumulated GPM associated with each section of pipe should be noted on the drawing.

Sizing Objectives

The size of the water pipe affects installation cost and pump operating cost. The pipe size also determines the water velocity, which is directly related to problems associated with cavitation and noise. Therefore, the water pipe sizes should be based on a design velocity or a design friction rate that will keep these factors within acceptable ranges.

Design Velocity

For hydronic systems, the water velocity design value can range between 1 and 10 FPS, but there is no specific recommended value. Some designers think that the design velocity should depend on the pipe size. For example, they may reduce the upper limit to 4 FPS when the pipe is two inches or smaller in diameter. (There are a number of opinions about the optimum design velocity, but there is not enough documentation to specify a specific value.)

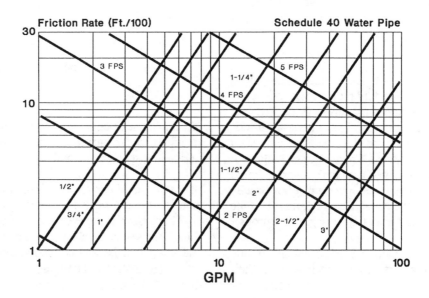

Figure A5-3

Design Friction Rate Value

The friction rate design value also is somewhat arbitrary. Designers typically use friction rate values between 1 and 4 feet of pressure loss per 100 feet of pipe. Any value that falls in this range is acceptable, providing that the corresponding velocity is acceptable.

Sizing

The size of each pipe section is determined by the flow rate (GPM) through the section, and the design friction rate (f/100) value. A pipe sizing friction chart (Figure A5-3) is required for this work. It is very important to use a friction chart that documents the performance of the material that will be used to fabricate the system. Friction charts are published for copper, steel, and plastic pipe.

Pressure Drop Calculations

An estimate of the system resistance must be made before the pump can be selected. Calculate the largest pressure loss associated with the supply runs and the largest pressure loss associated with the return runs. Include the pressure losses associated with the straight runs, fittings, valves, and other standard hardware items. Also include the pressure drops associated with the water side of the HVAC equipment.

The pressure drop associated with the straight runs, fittings, and the miscellaneous hardware depends on the friction rate and the total effective length of pipe. Refer to piping manuals for tables that list equivalent lengths for the various types of pipe fittings, valves, and other incidental hardware. The pressure drops associated with the water-side components of the HVAC equipment and the pressure drops associated with control valves are documented in manufacturers' engineering literature.

Pump Selection

Pump manufacturers' performance charts provide the required performance information. These summaries relate GPM and pressure to impeller size and motor power, for a specific RPM. Use the design values for the system GPM and head to determine the required impeller size and motor horsepower. Figure A5-4 provides an example of pump performance data.

Expansion Tank

Hot water or chilled water system operating pressures must be maintained within certain limits during any operating condition, even when the system is idle. This is accomplished by installing an expansion tank. This tank controls system pressure by accommodating the volumetric expansions and contractions associated with continuously changing water temperatures. The size of the expansion tank depends on the volume (in gallons) of the hydronic system, the operating temperatures, and the operating pressures. Hydronic specialties manufacturers provide comprehensive sizing guidelines and installation instructions in their engineering literature. Piping design manuals provide information about the volumetric properties of pipes, and HVAC equipment manufacturers provide information about the holding capacity of the water-side devices.

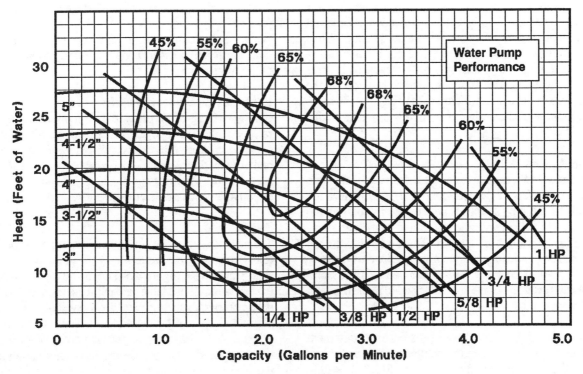

Figure A5-4

Hydronic Specialties and Detail Piping

As illustrated by Figure A5-5, pressure reducing valves, air separators, and air vents provide a few examples of important hydronic accessories. Hydronic specialties manufacturers provide comprehensive application, sizing, and installation information in their engineering literature. Pump, control valve, and HVAC equipment manufacturers provide information about the detail piping required for a particular device.

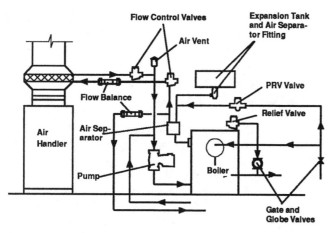

Figure A5-5

A5-3 Gas-Piping Systems

When a gas-piping system is designed, the pipe runs are sized so that the pressure drop from the point of delivery to the gas-burning equipment, is equal to, or less than, the value for the allowable pressure drop. This sizing procedure is based on the load condition that represents maximum demand. When there are multiple, nonconcurrent loads, the maximum demand value should be reduced by a percentage that reflects the effect of load diversity. If new loads will be added in the future, the maximum demand load should be adjusted upward. A brief overview of the design procedure is provided in this section.

Flow Rate Values

If the terminal flow rates are noted on a drawing of the pipe system, the cubic feet of gas per hour (CFH) flowing through any section of the pipe system is determined by totaling the CFH values associated with the mains and branches downstream from the section of interest. These calculations are made by starting at the ends of the pipe system and working back to the point of delivery. The accumulated CFH associated with each section of pipe should be added to the drawing of the pipe system.

The input CFH required for a piece of gas-burning equipment can be determined from the manufacturer's published specifications. If the gas input requirements are listed in heating capacity units (BTUH), the CFH input can be calculated by dividing the BTUH value by the heating potential of one cubic foot of gas.

Service Pressure

The pressure at the point of entry can vary from a few ounces to 50 pounds, depending on the operating characteristics of the local gas service. Low-pressure systems (0.50 PSIG and less) are normally used for residential applications, and higher pressures are used for commercial and industrial applications.

Pressure Regulators

The pressure at the entrance to the fuel burning equipment must be compatible with the manufacturer's specification. These pressures are usually smaller than the service pressure. One or more pressure regulators can be used to satisfy the manufacturer's requirements. The number of regulators that should be installed is related to the installation cost of the piping system.

- The system pressure can be reduced at the point of delivery, and a low-pressure piping system can be used to distribute gas to the equipment. The advantage to this approach is that only one regulator is required. However, if the loads are large or if there are many long runs, large pipes will be required and the installation cost can be high.

- High-pressure gas lines can be used for distribution, and the system pressure can be reduced at each piece of equipment. The advantage to this approach is that smaller pipes can be used, but each piece of equipment requires a regulator, which must be purchased and maintained.

Design Pressure Drop

For low-pressure gas systems (0.50 PSIG and less), the design pressure drop is normally 0.30 or 0.50 inches-water-column. When the service pressure is more than 0.50 inches-water-column, the design pressure drop is normally equal to 10 percent of the supply pressure.

Reference Equivalent Length

Gas-pipe sizing tables provide design information for a wide range of effective pipe lengths. However, when these tables are used, all the pipe runs associated with the system — no matter what their length — are sized by using a *single reference length*. This reference length is equal to the longest effective length associated with the system.

Sizing

The size of each pipe section is determined by the flow rate (CFH) through the section, and the reference length. Pipe sizing tables (see Figure A5-6 on the next page) are required for this work. These tables are published in sets. Each table in the set provides sizing data for a particular inlet pressure, pressure drop, and piping material. These tables are usually prepared for gas that has a specific gravity of 0.60. A correction is required if the specific gravity of the gas does not match the specific gravity that was used as the basis for the table. Specific gravity corrections are discussed in this section.

Gas Pipe Sizing Table — Pressure Under 1 Pound 0.50 Inch Water Column Pressure Drop 0.60 Specific Gravity; Schedule-40 Steel Pipe									
Pipe Size (Inches)	Total Equivalent Length of Pipe in Feet								
	50	100	150	200	250	300	400	500	1000
1.00	284	195	157	134	119	108	92	82	56
1.25	583	400	322	275	244	221	189	168	115
1.50	873	600	482	412	366	331	283	251	173
2.00	1681	1156	928	794	704	638	546	484	333
2.50	2680	1842	1479	1266	1122	1017	870	771	530
3.00	4738	3256	2615	2238	1983	1797	1538	1363	937
3.50	6937	4767	3828	3277	2904	2631	2252	1996	1372
4.00	9663	6641	5333	4565	4046	3666	3137	2780	1911

Figure A5-6

The pipe size for any pipe section is determined by entering the table at the reference length value, and reading down to the desired CFH value, then left to the pipe size. As emphasized earlier, the same reference length value is used for all the pipe sections associated with the system.

Specific Gravity Correction
The specific gravity of natural gas is about 0.60. Gases that have a lower specific gravity occupy a larger volume for a given weight (typically the volume associated with 1 pound of gas). In other words, gases with a lower specific gravity are less dense. Therefore, when the specific gravity is less than 0.60, larger pipes must be used to transport a given mass flow (pounds per hour). Conversely, smaller pipes can be used when the specific gravity of the gas is greater than 0.60. A table of specific gravity correction factors is normally provided with the gas-pipe sizing tables.

Liquid Petroleum Gas
The gas-piping tables can be used for sizing pipe for LPG systems. In this case, the service pressure will be determined by the pressure regulator at the storage tank. Also, note that liquid petroleum gas has a considerably higher specific gravity than natural gas.

Fuel-Air Mixtures
Air can be mixed with gas to control the heat content of the gas. Air-gas mixers are commonly used to make LPG gas behave like natural gas. This allows either fuel to be used without changing the burner setup. Since the LPG fuel-air mixture has a higher specific gravity than natural gas, the distribution piping is designed for natural gas.

Codes and Regulations
Operating pressures, sizing procedures, and stipulations about methods and materials are extensively codified. Make sure that the design conforms to building codes, fire codes, utility regulations, and insurance underwriter requirements.

A5-4 Oil-Piping Systems

Oil-piping systems can be classified as pressurized or unpressurized. Pressurized systems are used if the fuel oil has to be distributed to many burners (which may be scattered over a large area) or to a few, remotely located burners. Unpressurized systems can be used if one, or a few, burners are located close to the oil tank. A complete oil-piping system could consist of one or more burner pumps, a distribution pump, an oil tank, interconnecting piping, and various accessories such as filters, valves, gauges, and air vents.

If the oil distribution system is not pressurized (typical residential application), the burner pump draws oil directly from the storage tank. The simplest unpressurized system consists of a single suction pipe that connects a single-stage fuel oil pump to the oil storage tank. This arrangement is normally used when the oil storage tank is above the burner pump, but it can be used (providing that it is airtight) if the lift, associated with a tank that is installed below the pump, is not too large.

A two-pipe system is normally used when the fuel tank is located below the oil pump. In this case, the burner pump may be a single-stage, or a two-stage device, depending on the amount of lift required. If there is more than one oil burner, each burner pump will require a separate suction line, but a common pipe can be used for the return line. On the next page, Figure A5-7 provides examples of these two systems.

Flow Rate Values
If the terminal flow rates are noted on a sketch of the piping system, the gallons of oil per hour (GPH) flowing through any

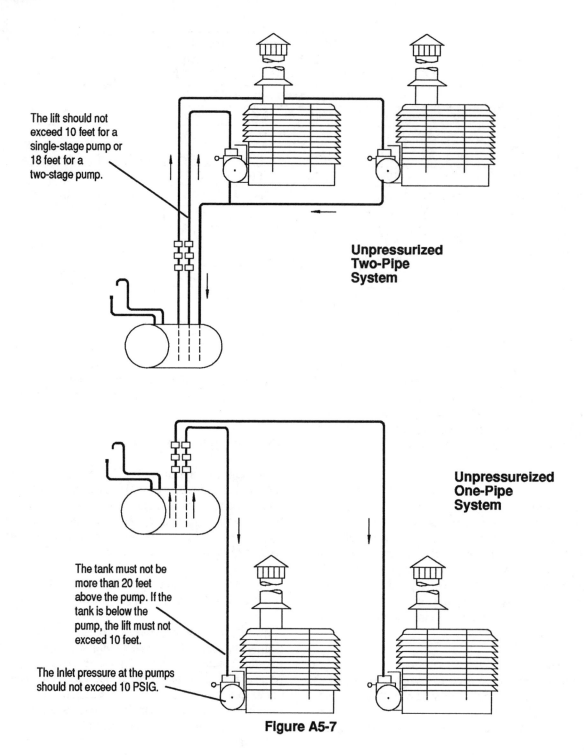

The lift should not exceed 10 feet for a single-stage pump or 18 feet for a two-stage pump.

Unpressurized Two-Pipe System

Unpressureized One-Pipe System

The tank must not be more than 20 feet above the pump. If the tank is below the pump, the lift must not exceed 10 feet.

The Inlet pressure at the pumps should not exceed 10 PSIG.

Figure A5-7

section of the system can be determined by summing the GPH values associated with the mains and branches downstream from the section of interest. These calculations are made by starting at the end (or ends) of the pipe system, and working back to the point of delivery. The accumulated GPH associated with an oil pump should be noted on the drawing.

The maximum oil flow rate required for a furnace can be determined from the manufacturer's specifications. If the input requirements are listed in heating capacity units (BTUH), the GPH input can be calculated by dividing the capacity value by the heating potential of a gallon of fuel.

Pump Sizing

The burner pump should be sized to deliver oil at the maximum firing rate, with some capacity held in reserve. The burner pump, or a burner pump specification, is normally provided with the heating equipment.

Distribution pumps are manufactured in different sizes, but the step from one size to the next size may be large. The distribution pump should be sized to deliver oil at the maximum combined firing rate associated with all the heating units, with some capacity held in reserve. When there are multiple, nonconcurrent loads, the maximum demand value should be reduced by a percentage that reflects the effect of load diversity. And, if new loads will be added in the future, the maximum demand load should be adjusted upward.

Pipe Size

The size of a suction pipe, distribution pipe, or return pipe is determined by tables that are (usually) provided by the pump manufacturer (see Figure A5-8). These tables list the maximum allowable pipe length that can be accommodated by a particular pipe size. This allowable length varies, depending on the performance of the pump and on the vertical lift associated with the run. (Pipe length calculations should include the equivalent lengths of the fittings, valves, and other hardware.) Different tables are required for different grades (weights) of oil.

Maximum Discharge Line Length (Feet)

Pump	GPH	1/2" Tube	1/2" Pipe	3/4" Pipe
P-30	30	300	800	2500
P-50	50	175	350	1500

Maximum 1/2" Suction Line Length (Feet)

Lift	0 - 7 Ft.	10 Ft.	13 Ft.	15 Ft.
P-30	100	80	63	52
P-50	60	53	41	34

Figure A5-8

Codes and Regulations

Sizing procedures and stipulations about methods and materials are usually codified. The details associated with locating, anchoring, and venting storage tanks also may be in the code. Make sure that the design conforms to building codes, fire codes, utility regulations, and insurance underwriter requirements.

A5-5 Steam-Piping Systems

Steam piping may consist of a single pipe, which carries steam and condensate; or two pipes, one for steam, and the other for condensate. The two-pipe system is commonly used for a wide range of heating applications. Two-pipe systems are further classified by five pressure categories:

- High-pressure — 100 PSIG and above
- Medium-pressure — 15 to 100 PSIG
- Low-pressure — 0 to 15 PSIG
- Vapor — vacuum to 15 PSIG
- Vacuum — vacuum to 15 PSIG

Two-pipe systems also are classified by the condensate return arrangement, which may be identified as a gravity return or a mechanical return. The gravity return can be used if all the heating units are located at a higher level than the boiler or condensate receiver water line. Mechanical return systems use steam traps, condensate pumps, or vacuum pumps to force the condensate back to the boiler.

Flow Rate Values

If the terminal flow rates are noted on a sketch of the pipe system, the pounds of steam per hour (LB/HR) flowing through any section of the system can be determined by summing the LB/HR values associated with the mains and branches downstream from the section of interest. These calculations are made by starting at the ends of the pipe system and working back to the point of delivery.

The input energy (LB/HR) required for a piece of heating equipment can be determined from the manufacturer's specifications. If the input requirements are listed in heating capacity units (BTUH), the LB/HR input can be calculated by dividing the BTUH value by the heating potential of a pound of steam.

Sizing Objectives

The sizing criterion consists of a design pressure drop for the supply side of the system, a design pressure drop for the return side of the system, a design value for the supply-side friction rate, and a design value for the return-side friction rate. Steam-piping manuals provide information about sizing criteria that should be applied to low-pressure steam systems and to other types of steam systems. The low-pressure sizing criteria is summarized below.

The steam-supply and condensate-return lines are sized so that the pressure drop associated with either side of the system does not exceed a recommended value. This recommended value depends on the operating pressure of the system. For example, in a low-pressure system, the design value for either side of the system can range from 0.50 to 5.0 PSIG, depending on the supply pressure at the boiler (2 to 20 PSIG).

The steam-supply and condensate-return lines also are sized so that the friction rate (pressure drop per 100 feet) associated with either side of the system does not exceed the maximum allowable value. Again, this allowable value depends on the operating pressure of the system. For example, in a low-pressure system, the supply-side friction rate should not exceed 2 PSIG per 100 feet, and the return-side friction rate should not exceed 0.50 PSIG per 100 feet.

Pressure Drop Calculations

Pressure drop calculations are based on the friction rate and the total effective length of pipe. Piping design manuals provide tables that document the equivalent lengths of fittings, valves, and other piping hardware.

Pipe Size

Steam-supply and condensate-return pipe sizes can be determined by using charts found in steam-system design manuals and piping handbooks. These charts can be used to estimate the friction rate (or pressure drop) associated with a given pipe size, based on the combination of supply pressure and steam flow rate.

Steam Trap Selection and Sizing

Steam traps are used on mains and risers, and they are installed downstream from heat transfer devices. Steam traps hold the steam in the supply side of the system, and to allow condensate and air to pass through to the return side of the system. The size of the steam trap depends on the load, the pressure difference across the trap, and the desired safety factor. A number of different steam trap designs are available in a wide range of sizes. Refer to steam trap manufacturers' engineering literature for detailed selection and sizing information.

Condensate Pump Sizing

Condensate pumps can be used to overcome the flow resistance associated with the return piping, the return risers, and the pressure difference between the return side of the system and the boiler. Refer to condensate pump manufacturers' engineering literature for selection and sizing information.

Piping Details

Many of the piping details associated with steam mains and runouts are related to condensate drainage and thermal expansion. The piping details associated with heat transfer equipment relate to vacuum release, and to protecting, isolating, and servicing the equipment. Refer to the manufacturer's literature and to steam piping handbooks for comprehensive information about piping details.

Codes and Regulations

Operating pressures, sizing procedures, piping details, and stipulations about methods and materials are extensively codified. Make sure that the piping system design conforms to all regulations.

Appendix 6
Venting Gas- and Oil-Heating Equipment

A vent is required for natural gas, LP gas, and oil-fired heating equipment. This vent must be designed so that all the combustion gases are expelled to the outdoors. Usually, these combustion gases must be discharged before condensation occurs in the vent system. This appendix summarizes the concepts and principles associated with venting gas- and oil-fired equipment.

A6-1 Terms and Definitions

There is a tendency to use the terms flue, stack, and vent interchangeably, but this is not consistent with industry practice. In this appendix, these terms are used in accordance with the following definitions.

- The term "vent" applies to any piping or chimney system used to expel the products of combustion to the outdoors. A vent could be under positive pressure, or it could be under negative pressure.

- The term "flue" applies to equipment that has a draft hood, draft diverter, or barometric. The flue pipe is located between the flue outlet and the draft hood, diverter, or barometric. Undiluted combustion gases flow through the flue pipe. If the equipment has a built-in draft diverter, there is no external flue pipe.

- The term "stack" applies to equipment that uses dilution air (equipment that has a draft hood, draft diverter, or a barometric). Diluted gases flow through the stack. Therefore, the stack is downstream from the device used to introduce the dilution air.

- The term "lateral" refers to the horizontal (or slightly inclined) portion of a single appliance vent system.

- The term "vent connector" refers to a pipe — which usually has a rise and an offset — that connects an appliance to a common vent.

A6-2 Factors That Affect Vent Performance

The vent design must consider the pressure inside the vent, the temperature of the vent gases, the ambient temperature surrounding the vent, the rate at which the combustion gases are generated, the conductivity of the vent walls, the thermal mass of the vent walls, the height of the vent, the length of the lateral(s), the vent gas dew point, and the condensate drain requirements.

In addition, a vent must terminate at a location that insures that the venting performance will not be affected by the building's structural components. And, the vent must terminate at a location that insures that the vented gases will not reenter the building through the fresh air intake, or through the windows or doors. Finally, adequate clearances must be provided between the venting system and the building components (or non-structural materials that are stored in the building).

Pressure in the Vent
The first thing to consider is whether the vent will operate at a positive or negative pressure. If the vent is under positive pressure, it must be air tight, or dangerous gases will be discharged into the building. The vent does not have to be air tight if it operates under a negative pressure. In fact, the potential for moisture condensation is slightly reduced if indoor air is drawn into the vent.

The pressure in the vent will depend on whether or not the combustion system is fan assisted. But, even if a fan is used, the vent may not be under positive pressure. Some fan-assisted combustion systems have powerful fans that pressurize everything downstream from the fan (mechanical draft); others, which only have enough fan power to move the combustion air through the burner and the heat exchanger, do not pressurize the vent.

If a fan forces air into the combustion chamber, it is referred to as a forced draft, power burner, power combustion, or pressure-fired system. If the fan exhausts the combustion chamber, it is referred to as an induced draft, induced vent, or power vent system. If a fan forces outside air into a sealed combustion chamber and forces the combustion gases out of the vent, it is referred to as a direct vent system.

If fan-assisted heating equipment pressurizes the vent, the equipment manufacturer will provide proprietary vent-sizing information. (In this case, the vent sizing problem is similar to a duct design problem — for a given rate of gas flow, the resistance of the vent must be matched to the external pressure produced by the fan.) The manufacturer also should provide information about the methods and materials required to install the vent. Depending on the temperature of the vent gas, the vent material could be plastic or metal. Of course, all the joints must be airtight when the vent operates under a positive pressure.

If fan-assisted heating equipment does not pressurize the vent (natural draft), or if no fan is used (atmospheric burner), the vent will operate under negative pressure, and the venting power will come from the stack effect. In this case, the manufacturer is still required to provide the necessary vent-sizing information, particularly if the product has a fan-assisted combustion system. However, it is possible that a manufacturer may not provide vent-sizing information. In this case, the National Fuel Gas Code (NFGC), the American Gas Association (AGA), the Gas Appliance Manufacturers Association (GAMA), the Hydronics Institute, and prefabricated vent and chimney manufacturers provide generic tables, which can be used to design the vent. More information about these tables is provided in Sections A6-7 and A6-8.

Some fan-assisted heating equipment may partially pressurize the vent. In this case, the manufacturer should provide specific instructions that explain how to design and install the venting system.

Vent Gas Temperature
The vent gas temperature is important. It affects the venting power of a natural draft vent, stack, or chimney, and it determines the potential for condensation to form in the vent. (Venting power increases and the potential for condensation decreases as the vent gas temperature increases.) The vent gas temperature also affects the type of material that can be used to build or fabricate the vent (see Section A6-3).

The temperature of the vent gas produced by gas-fired equipment is not as hot as the vent gas produced by oil-fired equipment (or solid fuel equipment). However, some newly developed oil-fired condensing furnaces and condensing boilers have relatively low vent-gas temperatures.

Approximate Vent-Gas Temperatures	
Gas-fired, condensing equipment	100 to 130°F
Gas-fired, mid-efficiency equipment, no draft hood	275 to 350°F
Natural gas, low efficiency equipment, with draft hood	360°F
LP gas, low efficiency equipment, with draft hood	360°F
Gas-fired, low efficiency equipment , no draft hood	460°F
Oil-fired, low efficiency equipment	560°F
Oil-fired, mid-efficiency equipment	400°F

Table A6-1

Besides the type of fuel, the vent gas temperature depends on the efficiency of the equipment, the amount of excess air, and the amount of dilution air (if the equipment has a draft hood, diverter, or barometric) that is mixed with the flue gas. Table A6-1 provides a few examples of the vent-gas temperatures associated with various types of equipment, but this table should not be used to design a vent system. Always refer to the manufacturer's installation literature for information about the vent-gas temperature associated with a particular piece of equipment.

Ambient Temperature
The ambient temperature surrounding the vent is important because it affects the venting power of a natural draft vent, stack, or chimney; and because it determines the potential for condensation to form in the vent. Venting power and the potential for condensation increases as the ambient temperature decreases.

Generation Rate
The rate at which the vent gases are generated depends on the input capacity (BTUH) of the burner. Therefore, instead of using vent-gas flow rate data (CFM or CFH); a vent, stack, or chimney is sized according to the input BTUH rating associated with a particular piece of equipment.

Conductivity of the Vent Wall
Wall losses, due to conduction, will cause the temperature of the vent gas to decrease as it moves through the vent. This temperature drop reduces venting power and increases the possibility that water vapor (or acid, in the case of oil-fired equipment) will condense in the vent. Therefore, double-wall vents and chimneys, which are designed to reduce the wall losses, are recommended for negative vent pressure, non-condensing, gas- and oil-fired equipment. (Manufactured double-wall chimneys, or masonry chimneys, are required if the vent-gas temperature exceeds the temperature limits associated with a B- or L-vent.)

Metal, single-wall vents are not recommended because they have relatively large conduction losses. They also require more clearance between the vent wall and any combustible material, including the building's structural components. However, if codes permit, they could be used if the designer is sure that condensation will not occur. But, this would depend on the initial temperature of the vent gas, the length of the vent, and the ambient temperature.

Thermal Mass of the Vent
The thermal mass of the vent is important because it affects the temperature of the vent gas during startup. If the walls have a large mass (masonry chimney), a considerable amount of heat will be extracted from the vent gas during the beginning of each operating cycle. This could cause moisture or acidic condensation to form in the vent, and it degrades venting power. Eventually, if the burner fires long enough, the vent might be heated to a suitable operating temperature, and the performance of the vent would be restored. However, it is possible — even probable — that a massive vent will not be heated to a suitable operating temperature when the burner

cycles to satisfy an off-peak load. (Part-load venting problems are exacerbated if the heating equipment is oversized.)

The vent walls will be cold during startup, so some condensation can be expected to occur in any type of vent system right after the burner begins to fire. This transient buildup of condensation is referred to as wet time. After a few minutes, the vent heats up, and this condensation usually disappears. The amount of condensation produced during the wet time, and the power of the drying action, depends on the type of equipment. Since mid-efficiency, gas- and oil-fired equipment have a longer wet time and less drying action, this equipment should not be vented into a chimney that has a lot of thermal mass, or a passageway that is surrounded by cold temperatures.

Metal vents and manufactured metal chimneys have an advantage over masonry chimneys because there is only a small amount of thermal mass associated with the metal walls. If a masonry chimney is used, it should have an air space between the tile liner and the masonry work. This way, the thermal mass associated with the tile liner is much smaller than the thermal mass of the masonry walls.

Even if a masonry chimney has an air space between the tile liner and the chimney walls, it may not be suitable for venting some types of equipment, because condensation problems could result if the temperature of the vent gas is too low. For example, moisture condensation could occur when a mid-efficiency, gas-fired, induced draft unit is vented through a masonry chimney, or acid condensation could occur if a mid-efficiency, oil-fired unit is vented through a masonry chimney. In these situations the chimney should be fitted with a double-wall liner.

Geometry of the Vent and the Lateral
The vent height is not a factor when the vent is pressurized, because the gases are forced through the vent by a fan. However, the flow resistance associated with the straight runs and the fittings is important. If there is too much resistance, the fan will not provide the required flow rate. Therefore, when the equipment pressurizes the vent, the equipment manufacturer must supply the information required to design a vent that is compatible with the fan.

The vent height is important if fan-assisted heating equipment does not pressurize the vent, or if there is no fan. In this case, the venting power depends on the height of the vent, stack, or chimney. (The venting power increases with the height.) Most natural draft installations require a vent that is at least 5 feet higher than the top of the equipment (always refer to manufacturer's instructions and to codes). Vented recessed wall heaters or wall furnaces require a vent that terminates at least 12 feet above the bottom of the heater (always refer to manufacturer's instructions and to codes).

The height of the vent is a primary factor that affects venting power of a single appliance vent system. However, if the gases from two or more pieces of natural draft equipment are vented into a common vent, the height of the connector risers are as important, or even more important, than the height of the common vent. (Combined venting situations are described in Section A6-6.)

The venting power of unpressurized vents also depends on the flow resistance associated with the straight runs and the fittings. If there is too much resistance, the vent will not produce the required venting power. Tables, which may be generic (NFGC, AGA, GAMA, Hydronics Institute, or prefabricated vent and chimney manufacturers) or product specific (published by the equipment manufacturer), must be used to design a negative pressure vent. More information about these tables is provided in Sections A6-7 and A6-8.

The resistance to the flow of the vented gas depends on the gas flow rate, the diameter of the vent, and the total equivalent length (*TEL*) associated with the straight runs and the fittings. Venting power of a vent, whether it is pressurized or not, can be increased by increasing the diameter, or by reducing the TEL of the vent. (But, at some point, the vent size becomes too large to establish a draft.)

A lateral, or a vent connector, is required when a vertical vent cannot be connected directly to the fitting located on the heating equipment. When a lateral or vent connector is required, it increases the resistance to the gas flow, and reduces the temperature of the vent gas. The flow resistance of a lateral or vent connector is accounted for by the tables used to size the vent system. The heat loss associated with the lateral or the vent connector walls can be minimized by using a double-wall pipe. (A double-wall pipe is especially important if a mid-efficiency, gas-fired, induced draft furnace is vented through a lateral or vent connector.)

In general, the horizontal run associated with a lateral or vent connector should be kept as short as possible, but the rise associated with a vent connector is important. (The equipment will not vent properly if the vent connector rise is too short, even if the common vent is very tall.)

Vent Gas Dew Point (Water Vapor)
The vent gas dew point (water vapor) depends on the type of fuel, the amount of excess air, and the amount of dilution air that is mixed with the flue gas. (Dilution air enters the vent through a hood, diverter, or barometric.) This temperature could range from less than 100°F, when natural gas is burned with a lot of excess and dilution air, to about 135°F, when

natural gas is burned with no excess air, and no dilution air. (The dew points associated with LP gas are lower than the dew points associated with natural gas.)

A low dew point temperature is desirable because it reduces the possibility of condensation. Therefore, the worst situation (limiting case) occurs when natural gas is burned with no surplus air (135°F dew point).

Vent Gas Dew Point (Acid)

With oil-fired equipment, there are two dew points — one is associated with the water vapor in the vent gas, the other is associated with the formation of acid. The acid dew point is the more important dew point, because acidic condensation occurs at a higher temperature than water vapor condensation, and because acid is more destructive than water.

The acid condensation dew point depends on the sulfur content of the oil. The acid dew point may fall between 250°F and 300°F if the oil has a high sulfur content, or it could fall between 225°F and 240°F if the oil has a low sulfur content. In either case, the acid dew point is not greatly affected by dilution air. Therefore, with oil-fired equipment, dilution air is not desirable because it lowers the stack gas temperature, but does not lower the acid dew point temperature.

Condensate Drain Requirements

Heat exchangers can be classified as non-condensing or condensing. A condensate drain will be required if condensation occurs in the heat exchanger. In this case, the drain is connected to the heating equipment. If no condensation occurs in the heat exchanger, a condensate drain may still be required, depending on the situation (see Table A6-2). In this case, the drain is connected to a trap built into the vent system. Refer to manufacturer's installation instructions for information about condensate drain requirements.

A6-3 Vent and Chimney Operating Temperatures

Plastic, single-wall vents can be used when a gas-tight, condensation resistant vent is required. PVC (Polyvinyl Chloride) can be used if the vent gas temperature is below 140°F, CPVC (Chlorinated Polyvinyl Chloride) can be used if the vent gas temperature is below 180°F, and PEI (Polyetherimid) can be used when the vent gas temperature is as high as 400°F. (Some venting system and equipment manufacturers offer proprietary vents that can withstand temperatures that approach 500°F.)

Metal vents are used when the vent gas temperature is in the 300°F to 500°F range. In this case, the vent could be a single-wall pipe or a double-wall pipe. Double-wall vents consist of two concentric metal pipes, which are separated by an air space. Type-B vents, which have a steel outer wall and an aluminum inner wall, are used only to vent gas-fired equipment. Type-L vents, which have a steel outer wall and a stainless steel inner wall, are designed for the temperatures

generated by oil-fired furnaces and boilers, but they also can be used for gas-fired equipment.

Manufactured, metal chimneys are designed to vent medium to high temperature gases (500°F to more than 2000°F). These chimneys also have a double wall, but in this case, an insulating material is sandwiched between the walls. These chimneys can be used to vent oil-fired equipment, and some codes may require a chimney for large gas-fired equipment. (They could be used in place of a B-vent or an L-vent, but the extra expense would not be justified.)

Masonry chimneys can withstand high vent gas temperatures, but they can cause problems when the vent gas temperature is moderate (less than 300°F), because they are more likely to cause the vent gas to condense. Masonry chimneys also should have a double-wall construction, with an insulating air space between the masonry walls and the tile liner.

Figure A6-1 summarizes the relationships between venting materials and venting applications. (Always use the vent pipe material recommended by the equipment manufacturer.)

Flue Materials		
Material	**Temperature**	**Application**
Various Types of Gas-Tight, Plastic and Metal Venting Materials	100 to 480°F	PVC, CPVC, PEI and proprietary products used to vent gas-fired equipment (1)
Single-Wall Metal Pipe	Refer to codes and to manufacturer's recommendations	Aluminum, galvanized steel, or stainless steel, depending on the application (2)
B- & BW-Vent	To 480°F	Gas equipment
L-Vent	To 500°F	Oil equipment (2)
Factory Built Chimney	500 to 2200°F	Oil or gas equipment (2)
Masonry Chimney (tile liner and air space)	360 to 1800°F	Gas or oil equipment (double-wall liner may be required for mid-efficiency equipment) (3)

1) Some high temperature plastic materials could be used to vent mid-efficiency (80 to 85 percent) oil-fired equipment if acidic condensation is expected or unavoidable.

2) Stainless steel vents can resist the heat that is associated with oil-fired vent-gases, but when the inner wall temperature exceeds 250°F, it loses its ability to resist acidic damage.

3) Masonry chimneys also are susceptible to acidic damage.

Figure A6-1

A6-4 Venting Requirements — Gas-Fired Equipment

Four venting categories have been assigned to gas-fired equipment — which could be a furnace or a boiler. The two factors that determine what category is assigned to a particular piece of equipment are the vent pressure and the equipment's efficiency. Another important consideration is whether the combustion system is designed to operate wet or dry. This information, and the vent system performance requirements associated with each venting category, are summarized in Table A6-2.

As indicated by Table A6-2, there are some venting problems associated with induced draft, mid-efficiency equipment that qualifies for Category 1, and some forced draft, mid-efficiency equipment that qualifies for Category III, based on the efficiency rating. (This equipment is characterized by a marginal difference between the vent gas temperature and the condensing temperature of the vent gas.) This type of mid-efficiency equipment is likely to produce condensation in the venting systems that have traditionally been used for higher temperature Category I and III equipment. (It seems that two more categories, perhaps I-A and III-A, could be created to

Gas-Fired Equipment Venting Categories		Category I	Category II	Category III	Category IV
Operating Characteristics (1992)	Pressure in the vent	Negative	Negative	Positive	Positive
	Annual efficiency	Below 83 Percent	Above 83 Percent	Below 83 Percent	Above 83 Percent
	Condensation	Not acceptable (Note 1)	Possible (in vent)	Possible (Note 3)	In heat exchanger
Design Requirements	Gas-tight vent	No	No	Yes	Yes
	Corrosion resistant vent	No (Note 1)	Yes	Possible (Note 3)	Yes
	Vent into masonry chimney	Permitted (Notes 1 & 2)	No	No	No
	Combined venting	Permitted	No	No	No
	Condensate drain	Not required (Note 1)	Ask manufacturer	Possible (Note 3)	At equipment
	Source of information regarding design and installation	Fuel gas code, heating equipment, and vent system manufacturers	Manufacturer's literature	Manufacturer's literature	Manufacturer's literature
Old Standard (Note 4)	Vent-gas dew point temp.	More than 140°F	Less than 140°F		
	Temperature of vent-gas (Note 5)	Above 275°F	Below 275°F		

Note 1 — Usually, there is no problem when high-vent-gas-temperature equipment is vented into a double-wall vent or into a lined masonry chimney; but condensation could occur if mid-efficency, (78 - 83%) mechanical draft equipment is vented into a vent that has highly conductive walls, cold walls, or massive walls. In this case, design a vent system that minimizes the wall losses (use double-wall pipe for the whole run and avoid long runs through cold spaces).

Note 2 — Install a B-Vent inside of the masonry chimney and use a double-wall connector when venting mid-efficency, (78 - 83%) mechanical draft equipment.

Note 3 — Condensation in the vent is possible with some types mid-efficiency (78 - 83%), direct-vent equipment, depending on the ambient temperature and the vent wall conductivity. In this case, design a vent system that minimizes the wall losses (use insulated pipe and avoid long runs through cold spaces). A corrosion resistant flue and a drain may be required if condensation cannot be prevented (refer to the manufacturer's recommendations).

Note 4 — This criteria is obsolete. Currently (1992), the criterion that is used to categorize the equipment is based on whether its efficiency rating is more or less than 83%.

Note 5 — The dew point of the vent gas depends on the fuel (natural or LP gas), the amount of excess air and the amount of dilution air. The limiting case occurs when the vent gas dew point is at a maximum, which is about 135°F. This maximum is produced when natural gas is burned with no excess air or dilution air. Therefore 275°F = 135°F dew point + 140°F rise.

Table A6-2

cover some types of condensation prone, mid-efficency, gas-fired equipment.)

Also note that the four gas equipment venting categories are usually associated with smaller, residential-type equipment (225,000 BTUH or less). This is because a wide variety of high-efficiency equipment is available in this capacity range. Most of the (larger) commercial equipment falls into Category I, which is covered by the National Fuel Gas Code. (Always refer to the manufacturer's instructions when installing Category II, III, and IV equipment, regardless of the heating capacity.)

A6-5 Venting Requirements — Oil-Fired Equipment

Depending on the burner design, oil-fired furnaces and boilers may require a natural draft vent or a vent that operates under positive pressure. If a natural draft vent is required, the manufacturer's installation instructions will provide a recommended connector size, vent size, and vent height. If a positive pressure vent is required, the manufacturer's installation instructions will provide the necessary design and installation information. (There also is some natural draft and mechanical draft vent-sizing information published in the sixth edition of the Hydronics Institute's boiler test and rating standard.)

The four venting categories assigned to gas-fired equipment do not apply to oil-fired equipment. (Oil-fired equipment may be categorized in the near future.) But, as noted above, oil-fired equipment could be classified by the pressure in the vent, which could be negative or positive.

Low-efficiency oil-fired equipment produces relatively hot vent gas temperatures — around 500°F — so the water vapor condensation problems associated with gas-fired equipment usually do not apply to oil-fired furnaces and boilers. However, if the vent wall temperature should drop below the acid dew point temperature, the vent will still be vulnerable to acid damage, regardless of equipment type.

Mid-efficency, oil-fired equipment (80 to 85 percent) produces cooler vent gas temperatures — around 400°F — so the water vapor condensation problems associated with gas-fired equipment could apply to this type of furnace or boiler. And, with this amount of vent gas temperature, the possibility of vent damage — due to acidic condensation — is more likely to occur.

Natural draft vents might be built out of single-wall steel pipe, Type-L vent pipe, or a masonry material; or they could be a steel, factory-built chimney (refer to the equipment manufacturer's instructions, and local and national codes). Positive pressure vents, which must be gas-tight, should be built in accordance with manufacturer's installation instructions and local codes. In either case, the vent must be designed to prevent acidic condensation, or it must be made of materials that will not be damaged by acid.

Acidic condensation may be prevented by using well-insulated, double-wall chimneys. (L-vents, which only have an air space between the walls, may not be adequate.) Also, try to avoid outdoor installations and long runs through cold spaces. If the vent has to be routed through a cold space, it should have enough insulation to prevent acid condensation. If acid condensation is possible, the vent can be formed out of an acid-resistant material. (If the vent gas temperature is below 480°F, a plastic vent material could be used, subject to UL approval.)

A few manufacturers offer high-efficiency oil-fired equipment (90 percent or more), which produces even lower vent gas temperatures. In this case, the equipment should be vented according to the manufacturer's installation instructions. (Plastic pipe is suitable for venting this type of equipment.)

A6-6 Combined Venting

A gas water heater, and Category I gas-fired heating equipment, can be vented into a common vent. In fact, when mid-efficiency heating equipment is used, combined venting is desirable. With this arrangement, the water heater will make the vent warmer, and it will provide some dilution air. (Dilution air lowers the dew point of the vent gas). Therefore, when a water heater is vented with a mid-efficency heating unit, the potential for condensation problems is reduced.

It also is acceptable to vent multiple, Category I, gas-fired heating units into the same vent. If all the units are in the same room, they can be vented through a manifold. If the units are located on different floors, they can be vented into a common vent, stack, or chimney. (In multi-story venting situations, the various pieces of equipment shall be installed in rooms that are separate from the occupied areas; and these equipment rooms shall be supplied with a source of outdoor air.) Information about designing these types of vent systems is published in the **National Fuel Gas Code,** and in booklets that are provided by vent and chimney system manufacturers.

The vent that is used with Category II, III, and IV gas-fired equipment should not be used by any other equipment (no combined venting). However, it may be permissible to route this type of vent through a chimney that is used for another appliance, but it must be so noted by the equipment manufacturer and permitted by the applicable codes.

Two or more oil-fired units can be vented into the same stack or chimney. Refer to manufacturer's installation instructions for the information required to design the manifold and breaching.

A6-7 Vent Sizing — Gas Equipment

Single appliance vent sizes depend on the material and construction techniques used to build the vent, the BTUH capacity of the equipment, the height of the vent, and the length of the lateral. Multiple appliance vent sizes (including manifold sizes and multi-story vent sizes) depend on the type of material and construction techniques used to build the vent, the BTUH capacity of the equipment, the height of the vent, and the height of the vent connector rise.

Tables that can be used to size vents for Category I gas-fired (atmospheric or fan-assisted natural draft) equipment can be found in the **National Fuel Gas Code**, the American Gas Association (AGA) venting tables, the Gas Appliance Manufacturers Association (GAMA) venting tables, booklets published by factory-built vent and chimney system manufacturers, and (possibly) the heating equipment manufacturer's installation instructions. These sources contain one set of tables that can be used to size single-appliance vents and a second set of tables that can be used to size vents serving two or more appliances (combined venting). Information about manifold venting and multi-story venting also may be included with the basic vent-sizing tables. Figure A6-2 summarizes the sources of information that pertains to the work associated with sizing Category I vents.

Category I Gas Vent Sizing Information (Atmospheric and Fan-Assisted)			
Type of Vent	NFGC AGA GAMA Tables	Vent-System Manufacturers	ASHRAE Handbook
Single-Wall Metal Vent	SA CV		Theoretical equations and charts that can be used to design masonry or metal chimneys, stacks, and vents.
Double-Wall Metal Vent	SA CV	SA CV	
Masonry Chimney	SA CV		
Factory-Built Chimney		Yes	
Manifold Venting		Yes	
Multi-story Vent Design	Yes	Yes	
Combustion Air Supply	Yes	Yes	Yes
Materials and Clearances	Yes	Clearances	Yes

1) SA = Single appliance; CV = Combined venting
2) Do not use for Category II, III, or IV
3) Do not use these tables if vent is under positive pressure
4) These vent-sizing tables can be used to find the maximum and the minimum size. Do not oversize vents (especially when venting mid-efficiency equipment, which typically has lower vent-gas temperatures and lower vent-gas flow rates than older, less efficient equipment.

Figure A6-2

Information about sizing vents for Category II, III, and IV gas-fired equipment is proprietary. Refer to the manufacturer's installation instructions for this information.

The four, gas-fired, equipment venting categories are usually associated with residential size equipment. However, the same principles apply to all sizes of gas-fired equipment, even if the category is not included on the manufacturer's nameplate label. In this case, the generic tables can be used to size vents for natural draft equipment; but proprietary tables, which should be supplied by the manufacturer, should be used to size positive pressure vents.

A6-8 Vent Sizing — Oil-Fired Equipment

Oil-fired equipment may operate with negative or positive pressure in the vent. Information about sizing vents for oil-fired equipment can usually be found in the installation instructions published by the heating equipment manufacturer. Also refer to the *June 1989 Test and Rating Standard for Heating Boilers, Sixth Edition,* which is published by the Hydronics Institute. For equipment that requires a natural draft, sizing and design information also can be found in the manuals published by L-vent, and factory-built, chimney system manufacturers. (There is an obvious need for more information about venting oil-fired equipment. Some useful information is being generated by researchers at the Brookheaven National Laboratory. This information will be published on a continuing basis.)

A6-9 Methods, Materials, and Clearances

Generic information about installation methods, materials, and clearances can be found in the **National Fuel Gas Code**, the **Uniform Mechanical Code**, in handbooks published by the American Gas Association and the Gas Appliance Manufacturers Association, the *ASHRAE Equipment Handbook,* and in vent-system manufacturers' manuals. Product specific information is provided in the heating equipment manufacturer's installation instructions.

A6-10 Combustion Air

An adequate source of combustion air is required. If indoor air is used as the source of combustion air, building codes usually specify the combustion air opening requirements. If there is no codified guidance, refer to the sources mentioned in the preceding paragraph.

A6-11 Draft Pressure Controls

Some types of heating equipment require a device that can be used to control the draft pressure. Refer to the manufacturer's installation instructions for information about this subject.

A6-12 Codes and Regulations

Local codes, regulations, and insurance underwriters' specifications establish the vent system design and installation requirements. These documents may simply defer to national codes and standards, or they may adopt the national codes with some revisions, or they may be proprietary codes that were developed by the local authority. In any case, the vent system should be installed in accordance with the local codes. If there is a conflict between the local code and the manufacturer's installation instructions, it should be brought to the attention of the local authority.

Appendix 7
Power Wiring

The electrical wiring and circuit protection for comfort-conditioning equipment must be able to satisfy the normal operating loads and the transients associated with starting the equipment. This wiring also must comply with applicable building codes and utility regulations.

A7-1 Power Requirements

Voltage, amperage, and power requirements are listed on the manufacturer's name/data plate, in published engineering data, and in the installation instructions provided with the equipment. The available power supply must be compatible with these requirements, which means that the service voltage and frequency must be within 10 percent of listed voltage requirement, and 3 percent of frequency requirement. In addition, the capacity (amperage) of the service, and the capacity of any circuit, must be adequate for the intended duty. If there is any question about the suitability of the power supply, consult with a power company representative.

A7-2 Internal Wiring

All wiring within the cabinet of manufactured products has been inspected during the rating, certification, and seal-of-approval process. This wiring should not be modified in the field, except by instruction of the manufacturer.

A7-3 Wiring Diagram

Internal wiring arrangements are normally documented in the diagrams provided with the manufacturer's engineering literature and installation instructions. A sketch of the external wiring should be provided by the installer and added to this reference material.

A7-4 Utility Regulations

Many utilities exercise oversight on issues that are tangentially related to providing metered electrical service. In this regard, a utility regulation may require documentation that demonstrates that the comfort-conditioning equipment is appropriately sized and properly installed, and that the conduction and leakage losses associated with the air distribution system are within an acceptable range. Utilities also may have incentive programs (cash rebates, reduced consumption charges, and low-interest loans) that encourage homeowners to install comfort systems that feature above-average efficiency. (These inducements should justify purchasing decisions that would otherwise have an unacceptably long payback period for the homeowner. In this regard, cash incentives paid to the contractor do nothing to improve the rate-of-return for the homeowner. If the concept being promoted has a relatively long payback period, the rebate should go to the homeowner.)

Appendix 8
Codes

Codes and ordinances provide a system of principles, guidelines, and rules that define minimum performance standards for the construction industry. Since these documents are statutory, designers and installers must be familiar with the requirements of every project.

A8-1 Specification Codes and Performance Codes

A code can be generally classified as a specification code or a performance code. Specification codes itemize requirements pertaining to design detail, materials, and construction techniques. Compliance with this type of code is reduced to demonstrating conformance with this list of mandates.

Performance codes establish minimum requirements for a performance-related goal. Therefore, there is no need for specific direction pertaining to design detail, materials, and construction methods. However, compliance with this type of code requires computational effort, because the contractor must demonstrate that the performance of the proposed design will be equivalent to, or better than, a specified benchmark or a codified surrogate design. Therefore, the code also must provide guidance regarding the analytical methods and software programs (calculation tools) that can be used to demonstrate compliance.

A8-2 Health Codes, Safety Codes, and Policy Codes

Codes that pertain to the integrity of the structural, electrical, plumbing, and sanitary systems were developed to protect the heath and safety of the citizenry. These documents are usually formatted as specification codes. Codes pertaining to energy efficiency convert policy generated by one or more government agencies into a statutory requirement. These documents take the form of a specification code or performance code, or the statute may option the use of either compliance path.

A8-3 Code Promulgation

Codes are promulgated by local, regional, state, and national authority. These documents are usually based on standards, manuals, test data, and other information generated by industry representatives (engineering societies, trade associations, manufacturer associations, and others), government agencies (the Departments of Energy, Housing and Defense, or the Federal Housing Authority, for example), and authorized laboratories and agencies (Underwriters Laboratory, the National Fire Protection Agency, insurance underwriters, and others).

In addition, contractors must conform to regulatory mandates generated by government agencies. In a legal sense, these requirements are not codes, but compliance is mandatory if the project is funded or supervised by government authority, or by an authorized agent of that authority. For example, the requirements associated with a FHA-financed home, or a home built on a military base, might not apply to an identical, privately financed home.

Since the responsibility for preparing and enforcing codes is widely distributed, jurisdictions may overlap; when there is a lack uniformity, the contractor may be subject to conflicting mandates. Confusion also occurs when agents at the field-inspection level are forced to interpret the intent of a code. When these situations occur, they should be brought to the attention of the jurisdictional authority; and to any other association or political agent that could assist with resolving the problem.

A8-4 Code-Writing Organizations

A partial list of organizations that create the standards that form the basis for building codes is provided here. However, the documents published by these groups do not have legal standing until they are adopted into law — in total or part — by a legislative process. In some cases, an equivalent document may be created by agents that are directly employed by a state or local government; or these agents may modify a published standard. In this regard, do not assume that familiarity with a published standard is equivalent to knowledge of a local code.

- The Building Officials and Code Administrators International (BOCA) provide standards for the Northeast states.

- The International Conference of Building Officials (ICBO) provides standards for the Midwest, Northwest, and Southwest parts of the country.

- The Southern Building Code Congress International (SBCCI) provides standards for most of the South.

- The National Electrical Code (NEC) sets standards for electrical wiring.

- The Air Conditioning Contractors of America (ACCA) publishes design manuals that are adopted into codes.

- The American Society of Heating, Refrigeration, and Air-Conditioning Engineers (ASHRAE) produces standards that pertain to testing and certifying HVAC equipment, and standards that pertain to energy efficiency as they apply to buildings and equipment.

- The Air Conditioning and Refrigerating Institute (ARI) sets standards for rating and testing unitary air conditioning equipment and heat pump equipment.

- The Gas Appliance Manufacturers Association (GAMA) sets standards for rating, testing, and venting furnaces and boilers.

- The American Gas Association (AGA) sets standards for rating, testing, and venting furnaces and boilers.

- The Underwriters Laboratories (UL) set standards for the electrical equipment and devices.

- The National Fire Protection Association (NFPA) sets standards of safety for structures, equipment, and distribution systems.

- The American Insurance Association (AIA) sets standards for fuel oil burning equipment, fuel oil storage, and piping.

- The Association of Home Appliance Manufacturers (AHAM) set standards for the room air conditioners and heat pumps (window units).

Index by Page Number